T0262184

Environmental Contamination Handbook

Environmental Contamination Handbook

Edited by **Emma Layer**

New York

Published by Callisto Reference,
106 Park Avenue, Suite 200,
New York, NY 10016, USA
www.callistoreference.com

Environmental Contamination Handbook
Edited by Emma Layer

International Standard Book Number: 978-1-63239-312-8 (Hardback)

Printed in the United States of America.

Contents

Preface

This book has been an outcome of determined endeavour from a group of educationists in the field. The primary objective was to involve a broad spectrum of professionals from diverse cultural background involved in the field for developing new researches. The book not only targets students but also scholars pursuing higher research for further enhancement of the theoretical and practical applications of the subject.

Environmental contamination is a topic of great concern. Environment reduces hazards, while man amplifies them. This is not a supposition, but a gist of the conclusions made from studies of experts from all over the world. The last two centuries have seen an unsystematic expansion and overexploitation of natural resources by man leading to damage and destruction of our own environment. Environmental pollution is the effect of illogical use of resources at the incorrect place and at the incorrect time. Environmental pollution has affected the lifestyle of people almost all over the world, and has decreased the span of life on earth. Today, we are forced to compromise with such environmental circumstances, which nobody had predicted would need to be tackled with for the survival of humanity and other life forms. This book will help the readers find out the underlying crisis and help monitor their solutions through various methods and approaches.

It was an honour to edit such a profound book and also a challenging task to compile and examine all the relevant data for accuracy and originality. I wish to acknowledge the efforts of the contributors for submitting such brilliant and diverse chapters in the field and for endlessly working for the completion of the book. Last, but not the least; I thank my family for being a constant source of support in all my research endeavours.

Editor

Part 1

Climate Change, Plants and Heavy Metal Contamination

Manganese: A New Emerging Contaminant in the Environment

Annalisa Pinsino[1,2], Valeria Matranga[2] and Maria Carmela Roccheri[1]
*[1]Dipartimento di Scienze e Tecnologie Molecolari e Biomolecolari
(Sez. Biologia Cellulare), Università di Palermo*
[2]Istituto di Biomedicina e Immunologia Molecolare "Alberto Monroy" CNR, Palermo
Italy

1. Introduction

The environment is composed of the atmosphere, earth and water. According to the World Health Organization, more than 100,000 chemicals are released into the global ambient every year as a consequence of their production, use and disposal. The fate of a chemical substance depends on its chemical application and physical-chemical properties, in combination with the characteristics of the environment where it is released. Chemical substances or contaminants discharged into the environment may be "natural" or "man-made". One of the most misunderstood concepts regarding contamination is the miss-interpretation of term "natural". A "natural" contaminant is one substance that can occur without human introduction. For example, trace metals, such as iron, zinc, manganese, copper, cobalt and nickel, can be considered naturally-occurring contaminants. Generally, these metals are found in the environment only in moderate amounts that do not cause health threats. However, "natural" contaminants can also have anthropogenic origins: in fact human activities often cause the release of a large amount of naturally-occurring minerals into the environment. Moreover, it is not the mere presence of a contaminant that makes it toxic, but its concentration. Paracelsus' famous aphorism "only dose makes the difference" has laid the groundwork for the development of the modern toxicology by recognizing the importance of the dose-response relationship.

In the last century, the massive production of manganese-containing compounds (metallurgic and chemical products, municipal wastewater discharges, sewage sludge, alloys, steel, iron, ceramics, fungicide products) has attracted the attention of scientists who investigated manganese as a potential emerging contaminant in the environment, and especially in the marine environment (CICAD 63, 2004). In humans, manganese excess is renowned for its role in neurotoxicity, associated with a characteristic syndrome called 'manganese madness' or 'Parkinson-like' diseases (Perl & Olanow, 2007). This neurodegenerative disorder is due to the accumulation of manganese inside intracellular compartments, such as the Golgi apparatus and mitochondria. In mammals, prenatal and postnatal exposure to manganese is associated with embryo-toxicity, fetal-toxicity, and decreased postnatal growth (Sanchez et al., 1993; Colomina et al., 1996).

In marine organisms some studied showed that an excessive amount of manganese causes toxicity, although the cause-effect evidence is not extensive.

In this chapter, we will provide: firstly the information available regarding the natural behaviour of manganese in the environment and its role in the living organisms, with particular emphasis on the marine environment. Secondly, we will discuss how and why the manganese contamination has become a global problem recently. Thirdly, we will cover some aspects regarding the adverse effects resulting from the exposure of whole organisms to high levels of manganese. Advantage of the notion that marine invertebrates express qualitatively similar types of induced damage to those found in higher organisms, we will focus our attention on the toxicity of manganese at different levels of organization: whole-organism, cellular and embryonic levels. In this review chapter we intend to promote the embryos and the immune cells of echinoderms as useful models to study manganese toxicity.

2. Manganese: Environmental aspects

Manganese is one of the most abundant and widely distributed metals in nature. In fact it is typically found in rocks, soils and waters. The Earth's crust consists of 0.1% of manganese.

As constituent of the soil, its concentrations range from 40 to 900 mg kg^{-1}. Pure manganese is a silver-stained metal; however, it does not occur in the environment in a pure form. Rather, it occurs in manganese-compounds, combined with other elements such as oxygen, sulphur, carbon, silicon and chlorine. These forms of manganese are solid and some of them can dissolve in water or be suspended in the air as small particles. The small dust particles in the air usually settle at the bottom within a few days, depending on their size, weight, density and weather conditions. Manganese can exist in 11 oxidation states, ranging from –3 to +7, but the most common ones are: +2 (e.g., $MnCl_2$) and +4 (e.g., MnO_2).

2.1 Manganese behaviour in the aquatic environment

Natural waters, such as lakes, streams, rivers and oceans, contain variable quantities of dissolved manganese, ranging from 10 to 10,000 µg l^{-1}. In water, most manganese compounds tend to attach to circulating particles or settle as sediment. Ocean spray, forest fires, vegetation, crustal rock and volcanic activity are the major natural atmospheric sources of manganese (CICAD 63, 2004).

Manganese exists in the aquatic environment in two main forms: Mn^{2+} and Mn^{4+}. Oscillation between these two forms occurs via oxidation and reduction reactions that may be abiotic or biotic (Schamphelaire et al., 2008). The interconversions between these forms is of particular importance to the aquatic chemistry of manganese, as Mn^{2+} forms are soluble whereas Mn^{4+} is present in insoluble oxides.

Since the late 1970s, the bacterially-catalyzed oxidation of manganese has been receiving increasing attention, because of its important role in geochemical cycles. Three well-studied and phylogenetically distinct manganese-oxidizing bacteria have been described: i) the β-Proteobacterium *Leptothrix discophora*, isolated from a swamp; ii) the γ-Proteobacterium *Pseudomonas putida*, which is a ubiquitous freshwater and soil bacterium; iii) the *Bacillus sp* spores, isolated from a near-shore manganese sediment (De Schamphelaire et al., 2007). A number of related metal-reducing micro-organisms have been identified and classified as the Geobacteraceae (Caccavo et al., 1994; Coates et al., 1995, 2001; Holmes et al., 2004; Vandieken et al., 2006). The biochemical pathways involved in Mn^{2+} oxidation have not yet been completely elucidated. A general scheme of the manganese cycle occurring in a sediment-water system is showed in Figure 1. The main oxidant in natural water is dissolved oxygen.

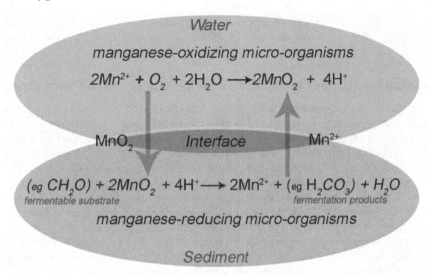

Fig. 1. A flux model for manganese interconversion. Manganese oxidation is performed in the oxic layer (water), while manganese reduction occurs in the anoxic layer (sediment). The oxic-anoxic boundary is located at the sediment-water interface. Note: all figures presented in this work are original.

The marine environmental chemistry of manganese is largely governed by pH, oxygen concentration of the solution and redox conditions. In fact, manganese oxidation increases with the decrease in acidity of the medium. The redox cycle of manganese in the oceans occurs at the oxic-anoxic boundary, which is often located at the sediment-water interface. Manganese oxides are present on the ocean floor as concretions, crusts and fine disseminations in sediments. It is well known, for example, that the soft bottom sediments of the oceans are particularly rich in manganese aggregates in the form of nodules (Bonatti & Nayudu, 1965; Wang et al., 2011).

Free manganese ions are released in the water by means of the photochemical and chemical reduction of manganese oxides coming from the organic matter (Sunda & Huntsman, 1998; De Schamphelaire et al., 2007). The process is initialised after the increase in temperature, the decrease in oxygen concentrations and the upward movement of the redox-cline (Balzer, 1982; Hunt, 1983). The transport of the dissolved manganese ions is governed by molecular diffusion in the water pores and it follows a manganese concentration gradient (the gradient decreases towards the oxic zone). In the marine environment, in absence of micro-organisms or mineral particles, manganese oxidation is a slow process (Wehrli et al., 1995). A reduced dissolved oxygen condition (called hypoxia) causes the rise of the ionic flux of manganese, which goes from the sediment to the overlying waters, where it reaches concentrations 1,000-folds higher than those normally occurring in seawater (up to 22 mg l^{-1}) (Trefry et al., 1984; Aller, 1994). Hypoxia in the marine environment can be natural or human-induced.

At present, costal hypoxia is increasing because of man-made alterations of coastal ecosystems and changes in oceanographic conditions due to global warming. In the deep ocean water, hypoxia is influenced mostly by the variations in the up-welling that is driven by the wind. Hypoxic areas are marine dead zones in the world's oceans which can happen for example in the fjords, coastlines or closed sea (such as Black, Baltic and Mediterranean Seas), where the water turnover, that should increase the oxygen content, is very slow or not present (Middelburg & Levin, 2009).

3. Biological functions of manganese in living organisms: General aspects

While manganese is abundant and widely distributed in nature, it is required only in trace amounts in the organisms during their life span, where it guides normal development and body function. In fact, it plays essential roles in many metabolic and non-metabolic regulatory functions, such as: i) bone mineralization; ii) connective tissue formation; iii) energetic metabolism; iv) enzyme activation; v) immunological and nervous system activities; vi) reproductive hormone regulation; vii) cellular defence; viii) amino acid, lipid, protein, and carbohydrate metabolisms; ix) glycosaminoglycans formation; x) blood clotting (ATSDR, 2008; Santamaria, 2008). Manganese works as a constituent of metallo-enzymes or as an enzyme activator. Examples of manganese-containing enzymes include: arginase, the cytosolic enzyme responsible for the urea formation; pyruvate carboxylase, the enzyme that catalyses the first step of the carbohydrate synthesis from pyruvate; and manganese-superoxide dismutase, the enzyme that catalyzes the dismutation of superoxide into oxygen and hydrogen peroxide (Wedler, 1994; Crowley et al., 2000). In contrast to the relatively few manganese metallo-enzymes, there are a large number of manganese-activated enzymes, including the hydrolases, the kinases, the decarboxylases, the DNA and RNA polymerases and the transferases (Missiaen et al., 2004). Activation of these enzymes can occur either as a direct consequence of the binding of manganese to the proteins, which causes subsequent conformational changes, or by its binding to the substrate, as in the case of ATP. Mechanisms regulating manganese homeostasis in cells are largely unknown. Some studies indicate the importance of regulated intracellular trafficking of manganese transporters to balance its absorption and secretion. Multiple transporters mediate intracellular manganese uptake including: i) natural resistance-associated macrophage proteins (Nramp); ii) cation/H+ antiporter; iii) zinc-regulated transporter/iron-regulated transporter (ZRT/IRT1)-related proteins (ZIP); iv) transferrin receptors; v) various calcium-transport ATPases; vi) glutamate ionotropic receptors (Au et al., 2008). Some of these transporters are localized within specific intracellular compartments, but none of them are manganese-specific transporters. In yeast, under normal conditions, the intracellular manganese concentration is regulated by adjustments of surface levels of the Nramp transporter, which regulates its degradation by endocytosis and ubiquitin-mediated targeting to vacuoles (Culotta et al., 2005). In mammals, the Golgi-associated secretory pathway Ca^{2+}-ATPase (SPCA) is known to pump cytosolic manganese into the lumen of the Golgi complex, in order to be used by glycosylation pathway enzymes. However, SPCA role in manganese detoxification has not been well elucidated (Missiaen et al., 2004).

4. Manganese toxicity: Causes and concerns

Manganese is considered an emerging contaminant because it is a perceived or real threat to the human health and the environment. Manganese exposure occurs at different levels and through a wide variety of industrial sources such as mining, alloy production, goods processing, iron-manganese operations, welding, agrochemical production and other anthropogenic activities. Manganese products can be discharged into the sea and become an unforeseen toxic metal in the marine environment. As manganese bioavailability increases, its uptake into living organisms occurs predominately through the water. The manganese rates of accumulation, as well as its elimination, are relatively fast-regulated processes. The exposure to high levels of manganese causes toxicity and decreases the fitness of the organisms (Roth, 2006). In humans, the neurological damage induced by excessive manganese exposure has been well documented for over a century (Cooper, 1837; Mena et al., 1967; Normandin & Hazel, 2002; Takeda, 2003). On the contrary, data on the effects of high manganese exposure in marine organisms are not well documented. In fact, although marine environment contains high natural concentrations of manganese, especially in the hypoxic zones, the potential danger to benthic and planktonic organisms has attracted the attention of scientists only recently.

4.1 Manganese toxicity in marine invertebrates

In all marine organisms manganese is accumulated into tissues; its amount reflects the concentrations of the bio-available manganese dissolved in sea water (Weinstein et al., 1992; Hansen & Bjerregaard, 1995; Baden & Eriksson, 2006). At the cellular level, manganese balance is proficiently managed by processes controlling cellular uptake, retention, and excretion (Roth, 2006), but these elaborate homeostatic mechanisms are altered under high levels of the available metal. Thus, it is important to consider that manganese dissolved in sea water is bio-concentrated significantly more at lower than at higher trophic levels (CICAD 63, 2004). The Bio Concentration Factor (BCF) correlates the concentration of a substance in animal tissues to the concentration of the same substance in the surrounding water. The reported BCF values range between: 100-600 for fish, 10,000-20,000 for marine and freshwater plants, 10,000-40,000 for invertebrates (ATSDR, 2008). In aquatic invertebrates manganese uptake significantly increases with temperature increase and salinity and with pH decrease. Dissolved oxygen has no significant effect (Baden et al., 1995).

Crustaceans and molluscs are the most manganese-sensitive invertebrates, followed by arthropods and echinoderms. The first studies on the effects of manganese in crustaceans (species *Homarus gammarus and H. vulgari*) were carried out by Bryan & Ward (1965).

High levels of manganese have been found in the haemolymph and body tissues of the lobster *Nephrops norvegicus* living in the SE Kattegat, Swedish west coast, as well as in lobsters living in the hypoxic areas near sludge dumping sites in the Firth of Clyde, Scotland (Baden & Neil, 1998). Exposure to high manganese impairs the lobster's antennular ficking activity, causing disorientation and inability to locate food (Krång & Rosenqvist, 2006). Likewise, unhealthy blue crabs, *Callinectes sapidus*, have been found in a manganese-contaminated area of North Carolina, USA (Gemperline et al., 1992; Weinstein et al., 1992). In general, internal tissues such as the intestine, nervous system, haemolymph and reproductive organs accumulate much more manganese than other tissues such as exoskeleton, but in the latter case, manganese elimination is a very slow process. In *Nephrops norvegicus*, manganese accumulation reached a plateau after 1.25 days of exposure in all tissues except for the mid-gut gland, which continued to accumulate manganese over time

(Baden et al., 1999). A similar accumulation pattern of manganese in soft tissues has also been described in mussels (Regoli et al., 1991). Specifically, in the species *Donacilla cornea*, manganese was rapidly accumulated; reaching a maximum after 3 days of exposure, and it was rapidly excreted (60% loss) after 3 days in clean sea water. Seasonal and sex differences in the manganese accumulation levels have been reported for both mussels (e.g. *Mytilus edulis* and *Mytilus californianus*) and oysters (e.g. *Crassostrea gigas* and *Crassostrea virginica*) (Nørum et al., 2005). For example, in the species *Mytilus edulis* the gonads of females accumulated manganese more than males. Manganese accumulation in the sea star *Asterias rubens* has shown linearity with time up to 23 days at low concentrations (0.1 mg l⁻¹), but its saturation kinetics were very fast at higher concentrations (Hansen & Bjerregaard, 1995). In fact, it was found that steady-state levels were reached in the coelomic fluids after only 5 days of exposure to 5.5 mg l⁻¹ (Oweson et al., 2008).

Manganese excess may cause a Ca^{2+} pump dysfunction, affecting neuro-muscular transmission in benthic marine invertebrates (Hagiwara & Takahashi, 1967; Baden & Neil, 1998; Holmes et al., 1999). For example, in crustaceans, manganese acts as a competitive inhibitor of the calcium-regulated ion channels present in nerve and muscle membranes, thus inhibiting synaptic and neuromuscular transmission and muscle excitation (Hagiwara & Takahashi, 1967; Holmes et al., 1999).

Manganese affects the immune system of marine invertebrates in a species-specific manner.

In the immune system of *Nephrops norvegicus* (haemolymph), manganese is mainly found in the protein fraction that includes haemocyanin and immune cells (called haemocytes). Recent studies showed that high levels of manganese affect *Nephrops norvegicus* haemocytes causing: i) apoptosis-induced reduction of the number of circulating haemocytes ; ii) inhibition of their maturation to granular haemocytes; iii) inhibition of the recruitment of haematopoietic stem cells (Hernroth et al., 2004; Oweson et al., 2006). These immune suppressive effects were also found in *Mytilus edulis* (Oweson et al., 2009). In addition, manganese alters the immune system of sponges (*Geodia cydonium*, *Crella elegans* and *Chondrosia reniformis*) by inhibiting the activity of the 2′, 5′-oligoadenylate synthetase (2-5A synthetase), an enzyme known to be involved in the functioning of the immune system of vertebrates (Saby et al., 2009).

Surprisingly, in contrast to what was recorded in crustaceans and molluscs, in echinoderms (*Asterias rubens*) manganese exposure stimulated haematopoiesis, thus causing an increase in the number of circulating immune cells (Oweson et al., 2008). Manganese effects on *Asterias rubens* immune system will be discussed in detail in the next sections.

4.2 How does manganese affect echinoderm immune cells?

Echinoderms play a key role in the maintenance of the integrity of the ecosystem where they live (Hereu et al., 2005) and are constantly exposed to pollutants deriving from different kinds of human activities (Bellas et al., 2008; Rosen et al., 2008). They are phylogenetically related to vertebrates and have a sophisticated and sensitive immune system. In echinoderms, immune cells (called coelomocytes) are a heterogeneous population of free moving cells found in all coelomic spaces, including the perivisceral coelomic cavities and the water-vascular system (reviewed in Matranga, 1996; Glinski & Jarosz, 2000; Smith et al., 2010). They are also present sparsely in the connective tissue (mesodermal stromal tissue) and amongst tissues of various organs (Muñoz-Chápuli et al., 2005; Pinsino et al., 2007). Coelomocytes participate as immune cells in function similar to their vertebrate's immune system homologues. In fact, they are involved in: clot formation, phagocytosis, encapsulation and clearance of pathogens, as well as oxygen transport. The coelomic fluid in which the immunocytes or coelomocytes reside and move is a key factor governing the immunological capabilities of echinoderms, as it contains essential trophic and activating factors (for a review see Matranga et al., 2005; Smith et al., 2010). Four different morphotypes have been described in the asteroid *Asterias rubens*, with the phagocytes as the most abundant type, accounting for approximately 95% of the total population (Pinsino et al., 2007).

As previously reported, the accumulation of manganese into the coelomic fluid of exposed sea stars (*Asterias rubens*) induces the proliferation of haematopoietic cells (Oweson et al., 2008). Specifically, by using the substitute nucleotide 5-bromo-2'-deoxyuridine (BrdU) for tracing cell division, and by recording the mitotic index after nuclei staining, authors found that manganese induced the proliferation of cells from a putative haematopoietic tissue, the coelomic epithelium. In addition, the haematopoietic tissue and coelomocytes showed stress response in terms of changes in HSP70 levels and protein carbonyls. Incubation with heat-killed FITC-labelled yeast cells (*Saccharomyces cerevisiae*) exhibited an inhibited phagocyte capacity of coelomocytes. Moreover, measurement of dehydrogenase activity, using MTS/PMS, revealed that manganese showed cytotoxic properties. Although manganese was revealed as stressful to the coelomocytes and affected their ability to phagocyte, the increased number of coelomocytes compensated these impairments. In summary, the authors concluded that the exposure of *Asterias rubens* to manganese impaired their immune response, but induced renewal of coelomocytes, assuring survival. Co-occurrence of manganese with hypoxic conditions does not inhibit the elevated production of coelomocytes, but probably affects the composition of the subpopulations of these immune cells since hypoxia, but not manganese, increased the mRNA expression of Runt, a transcription factor, assumed necessary for cell differentiation (Oweson et al., 2010).

5. Sea urchin embryonic development

To address this issue, at the beginning of this section we will describe the basic steps of the sea urchin development. Briefly, upon appropriate stimulation, millions of eggs and sperm are released into the sea water; after fertilization, the single-celled zygote is converted into a multi-cellular embryo through rapid and repeated mitotic cell divisions (cleavage). Founder cells of the three germ layers ecto- meso- and endoderm, are the basic units where regulatory information is localized during cleavage. In particular, β-catenin is required for the development of all endo-mesoderm territories, including the archenteron, the primary mesenchyme cells (PMCs) and the secondary mesenchyme cells (SMCs) (Logan et al., 1999). Cell fates are fully specified by the blastula-early gastrula stage of development, when cells have begun to express particular sets of territory-specific genes (Davidson et al., 1998). Although maternal determinants are required for founder cell specification during development, interactions between the PMCs and external cues derived from the ectoderm specify many phases of the skeleton formation (Armstrong et al., 1993; Ettensohn & Malinda, 1993; Guss & Ettensohn, 1997; Zito et al., 1998). The blastula stage is characterized by the presence of a large fluid-filled blastocoels, surrounded by a single layer of cells.

During gastrulation extensive cellular rearrangements occur which convert the hollow-spherical-blastula into a multi-layered gastrula. Changes in shape and differentiation of embryo structures lead to the formation of a pluteus, the first larval stage. Genus-specific spicule growth and patterning is completed at this stage, directed by the spatial-temporal regulated expression of bio-mineralization related genes (Zito et al., 2005; Matranga et al., 2011). Sea urchin development from the blastula to the pluteus stage is showed in figure 2.

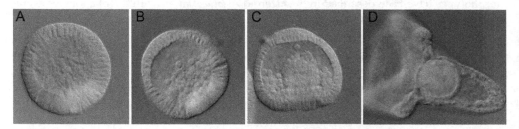

Fig. 2. Sea urchin development from the blastula to the pluteus stage. A) hatching blastula; B) mesenchyme blastula; C) middle gastrula; D) pluteus. Note: all figures presented in this work are original.

5.1 Sea urchin embryos as an *in vivo* model for the assessment of toxicity

The sea urchin is estimated to have 23,300 genes with representatives of nearly all vertebrate gene families (Sea Urchin Genome Sequencing Consortium, 2006). Since it has been demonstrated that the sea urchin genome shares at least 70% of the genes with the mankind, we shall consider how this provides an important tool kit to aid our understanding of eco-embryo- and geno-toxicological studies as well as for studies on embryonic development. Sea urchins are marine invertebrates with two life stages: i) an early and brief developmental stage (planktonic) and ii) a remarkably long-lived adult stage with life spans extending to over a century (epi-benthonic). Sea urchins are pivotal components of sub-tidal marine ecology (Hereu et al., 2005) and they are continuously exposed to environmental pressure, including changes in temperature, hypoxia, pathogens, UV radiation, free radicals, metals and toxicants. These marine invertebrates produce large numbers of susceptible, but not vulnerable, transparent embryos. The keys for their developmental success are the potent cellular mechanisms that provide them with protection, robustness and resistance, as well as the regulatory pathways that alter their developmental course in response to the conditions encountered (Hamdoun & Epel, 2007). The integrated network of genes, proteins and pathways that allow an organism to defend itself against chemical agents is known as the "chemical defensome" (Goldstone et al., 2006). In sea urchin embryos, many "defensome" genes are also expressed during their normal development as integral part of the developmental program, suggesting a dual function regulating both defence and development. In addition, genes involved in signal transduction often respond to environmental stress, activating alternative signalling pathways as a defence strategy for survival (Hamdoun & Epel, 2007). Thus, the sea urchin becomes an excellent candidate for the understanding of the two-fold function of genes/proteins and signalling pathways involved in both defence and regulation/preservation of development during environmental changes.

To date, several researchers have shown that exposure of sea urchin embryos to chemical and physical agents involve a selective set of defence "macromolecules" (Geraci et al., 2004; Roccheri et al., 2004; Bonaventura et al., 2005; Matranga et al., 2010; Russo et al., 2010; Pinsino et al., 2011a). Of interest, for example are reports about the biochemical and molecular changes occurring in response to cadmium exposure in *Paracentrotus lividus* embryos. Briefly, the toxic effects have been studied by examining the: accumulation, embryonic malformation, stress gene expression, stress protein induction, apoptosis and related pathways (Roccheri & Matranga, 2010). Specifically, it was found that the exposure of embryos to sub-acute/sub-lethal cadmium concentrations was able to trigger the expression of one of the metallothionein genes which binds metal ions in the cytoplasm for storage and/or detoxification (Russo et al., 2003). Simultaneously, or alternatively, cadmium was able to induce the new synthesis of several stress proteins—HSPs that usually facilitate the repair of miss-folded proteins or the elimination of aggregated proteins (Roccheri et al., 2004). The authors found that 9 hours of cadmium exposure were required to induce the synthesis of HSPs 70 and 72, while at least 15 hours were needed to observe the induction of hsp56 and 25kDa synthesis. In addition, it has been demonstrated that a long-lasting exposure (over 24 hours) triggers DNA fragmentation and causes the activation of caspase-3, one of the key molecules promoting apoptosis, which increased in a time-dependent way (Agnello et al. 2006, 2007). In sea urchin embryos, apoptosis is an important part of the defence strategy, both in physiological or stress conditions (Agnello & Roccheri, 2010).

Recently, it has also been demonstrated that in sea urchin embryos autophagy is a further defence strategy activated in response to cadmium exposure (Chiarelli et al., 2011). These authors found that autophagy reaches its maximum peak after 18 hours, when apoptosis is just beginning, suggesting that this degradation process starts before apoptosis and after the failure of HSP and metallothionein function. In conclusion, data demonstrate the wide range of alternative strategies that can occur at different levels of stress.

5.2 Manganese embryo-toxicity in sea urchin embryos: Biochemical and molecular studies

As previously mentioned, prenatal and postnatal exposure to manganese in mammals is associated with embryo-toxicity, fetal-toxicity, and decreased postnatal growth (Sanchez et al., 1993; Colomina et al., 1996; Doyle & Kapron, 2002; Giordano et al., 2009). Nevertheless, functional data on the effect of high manganese exposure on gene expression and on cellular mechanisms involved in embryonic development remain scant. Recently, we took advantage of the amenable embryonic model, the Mediterranean sea urchin *Paracentrotus lividus*, to investigate the potential toxicity of manganese on embryonic development, using different biological and biochemical approaches.

In our studies, embryos were continuously exposed to manganese from fertilization, at concentrations ranging from 1.0 to 61.6 mg l⁻¹ (or from 0.018 to 1,120 mM), and harvested at different developmental stages (Pinsino et al., 2010; Pinsino et al., 2011b). The biological study was carried out according to classical toxicological criteria, namely: concentration- and time-dependent responses, analysis of the impact on development, manganese accumulation. We found that embryos showed an elevated tolerance/resistance to manganese, as they accumulated high amounts into cells in a time- and concentration-dependent manner. Here we show, just as an example, the time course of manganese accumulation analyzed from 24 to 72 hours of exposure/development in embryos exposed to different manganese concentrations (Figure 3A).

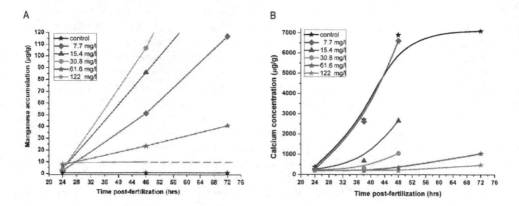

Fig. 3. Time course of manganese accumulation and calcium content determined by AAS in embryos exposed to different manganese concentrations (0, 1.0, 7.7, 15.4, 30.8, 61.6 and 122 mg l $^{-1}$). Note: all figures presented in this work are original.

Results were compared to the physiological calcium content measured in the same samples (Figure 3B). We found that calcium content diminished in an inversely proportional way to manganese accumulation. The amount of manganese accumulated and the calcium content in cells were determined, in exposed and control embryos, by atomic absorption spectrophotometer (AAS). Moreover, AAS data for calcium content was consistent with its poor detection in PMCs observed by *in vivo* labelling with the cell-permeable fluorescent dye, calcein (Pinsino et al., 2011b).

Rising manganese exposure concentrations from 1.0 to 61.6 or 122 mg l⁻¹ did not produce lethal effects. Rather, we observed a concentration-dependent increase in the number of morphological abnormalities found 48 hours post-fertilization (pluteus stage) (Figure 4). The impact on embryonic development was analysed considering that normal embryos should satisfy some morphological criteria as the correct schedule in reaching the developmental endpoint (pluteus) and the correct skeleton development and patterning (Pinsino et al., 2010). Major developmental defects consisted in the reduced elongation of skeletal rods (spicules) (see arrows Figure 4), suggesting a key role for manganese in embryonic skeleton development.

In addition, a correlation was observed when comparing malformations, accumulation of manganese and the regulation of key stress proteins that provide protection against stressors. In fact, embryos exposed to high manganese concentrations (15.4 mg l⁻¹ or above) showed an increase of the hsc70 and hsc60 protein levels at the 48 hours. The proteins are useful to protect them from apoptosis, in accordance to the finding that no DNA fragmentation was induced by manganese exposure (see Pinsino et al., 2010). By a fluorescent detection assay on live embryos, we found no induction of the reactive oxygen species (ROS), indicating no correlation between manganese toxicity and oxidative stress (see Pinsino et al., 2010).

In a recent work, we thoroughly extended our studies concerning the effects of manganese on skeleton development of *Paracentrotus lividus* embryos at the biochemical and molecular levels (Pinsino et al., 2011b).

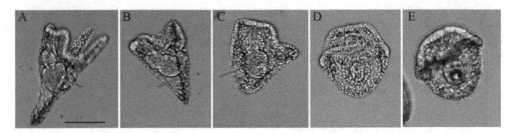

Fig. 4. Examples of manganese exposed embryos showing different phenotypes. A closer look revealed reduction and lack of skeleton elongation, depending on the manganese concentration used. A) control embryo; B) 7.7 mg l⁻¹; C) 15.4 mg l⁻¹; D) 30.8 mg l⁻¹; E) 61.6 mg l⁻¹. Arrows indicate skeletal rods. Note: all figures presented in this work are original.

To this purpose, we used the highest manganese concentration (61.6 mg l-1 or 1,120 mM) which we have previously demonstrated to prevent skeleton growth and produce spicule-lacking embryos. At first, we determined the effects of manganese exposure on the differentiation of the three germ layers (ecto- endo and mesoderm) by immuno-staining manganese-exposed embryos with UH2-95, 5C7, and 1D5 monoclonal antibodies (mAbs) which recognize antigens present on the ciliary band, midgut/hindgut, and PMCs. The three germ layers markers were detected at the appropriate time and in the correct position, confirming that they were at least expressed properly. We should remember that the PMCs are the only cells in the embryo that synthesize the protein components of the tri-radiate spicules and set the limits of calcite deposition. Despite the fact that no biomineral deposition was observed in exposed embryos, PMCs maintained the capacity to migrate and pattern inside the blastocoel's, as they did in control embryos, excluding the possibility that the lack of skeleton formation was caused by the PMCs miss-localization. Rather, PMCs showed a strong depletion of calcium in the Golgi regions, suggesting that manganese competes with calcium uptake and internalization (Pinsino et al., 2011b).

By *in situ hybridization,* we analyzed the expression of three genes expressed during PMC differentiation: *Pl*-sm50, *Pl*-sm30 and *Pl*-msp130, encoding two spicule matrix proteins and the cell surface protein detected by the 1D5 mAb. Results showed that in manganese exposed embryos: i) *Pl*-sm50 expression was largely normal; ii) *Pl*-msp130 expression was not down-regulated during development (compare Figure 5A, 5A', with Figure 5B, 5B') and iii) *Pl*-sm30 expression was severely reduced (compare Figure 5C, 5C', with Figure 5D, 5D').

As is well known, the three gene products participate in the synthesis of the skeleton, but the function of each of them it is not well understood yet. Since *Pl*-msp130 remains expressed over time in all PMCs of the embryos exposed to manganese, our data reinforces the idea that this cell-surface glycoprotein is directly involved in the control of the nucleation during solid-phase crystallization. Instead, *Pl*-sm30 protein seems to lead the elongation phase as supported by the down-regulation of its transcript over time (Figure 4), in agreement with reports on the American species *Lythechinus pictus* (Guss & Ettenhson, 1997).

It has been widely demonstrated that Extracellular signal-Regulated Kinase (ERK) MapK mediated signalling controls the expression of several regulatory genes which participate in the specification and differentiation of the sea urchin skeleton (Fernandez-Serra et al., 2004; Röttinger et al., 2004).

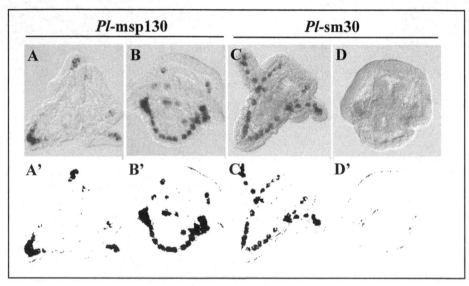

Fig. 5. Whole-mount in situ hybridizations in control (A, A', C, C') and manganese-exposed embryos (B, B', D, D'), performed with the following probes: *Pl*-msp130 (A-B, A'-B') encoding for PMCs surface protein; *Pl*-sm30 (C-D, C'-D') encoding for an integral spicule matrix protein.Note: all figures presented in this work are original.

Thus, we analyzed, by Western blotting, the activation of ERK in manganese exposed embryos during development. We found a persistent phosphorylated state at all stages examined, as proteins levels were only partially modulated during development of exposed embryos, contrary to the physiological oscillations observed in normal embryos.

In conclusion, our results showed for the first time the ability of manganese to interfere with calcium uptake and internalisation into PMCs, and the involvement of endogenous calcium content in regulating the activation/inactivation of ERK during the sea urchin embryo morphogenesis (Pinsino et al., 2011b).

6. Conclusion

Metals are one of the most abundant classes of contaminants generated by human activities and represent an actual hazard for marine ecosystems and organisms' health. In fact, although metals are terrestrially produced, they flow into the sea through effluent and sewage or are directly discharged from industries placed on the sea water front. Marine organisms can take up from the sea and from their diet these metals, which may consist of particles in suspension or might be deposited in the sediment. A great number of factors may influence dose-effect and dose-response relationships between metals and organisms; their tolerance to and use of trace metals reflect sea water concentrations. Marine invertebrates accumulate and bio-concentrate metals more than higher organisms; thus, their peculiar position in the marine trophic chain, where pelagic larvae are part of the diet of several planktonic and benthic organisms (bio-magnification), increases the interest of many researchers. Trace essential metals are of environmental interest both as limiting nutrients (Fe, Zn, Mn, Cu, Co, Mo and Ni), playing important roles in metal-requiring and metal-activated enzyme systems, and as toxicants when present at high concentrations. On the contrary, non-essential heavy metals as Cd, Hg, Ag, Pb, Sn and Cr, are toxic for living organisms even at low concentrations.

The effects of several metals on echinoderm embryos have been studied for many years. The response of sea urchin embryos to metal exposure involves selective sets of defence "macromolecules". The type of defence response elicited greatly depends on the different sensitivity of the organisms to the different metal used. Metal toxicity can trigger several biochemical and cellular events that include: the induction of a set of highly conserved proteins, such as the heat-shock proteins and the metallothioneins; the efflux transporters; the activation of anti-oxidative enzymes; autophagy and apoptosis. In this chapter it was reported that manganese-exposed embryos do not activate a fast defence response, but activate or repress signalling pathways and transcriptional activities involved in the regulation/preservation of development. This surprising behaviour is probably due to the fact that a moderate manganese increase is not recognized by the cell as stressful event. In fact, low manganese concentrations stimulate embryos growth (not shown), probably as a consequence of their role in the metabolic-enzymatic reactions. On the contrary, non-essential heavy metals, such as cadmium, activate several cellular defence strategies aimed at permitting embryo survival. In fact, *Paracentrotus lividus* embryos showed an elevated tolerance to manganese, due to the increased HSC levels, did not activate the synthesis of the HSPs inducible forms and did not enter into the apoptotic program. However, manganese interferes with calcium uptake and it's internalization into embryos, suggesting that skeletal growth is highly dependent on calcium signalling. The use of manganese-exposed embryos as a new model to study signalling pathways involved in skeletogenesis provides new insights into the mechanisms involved in manganese embryo-toxicity and emphasizes the role of calcium trafficking, recruitment and storage in the bio-mineralization process.

Interestingly, recent studies suggested that the homeostatic mechanisms that control the amount of trace metals necessary to metabolic activities are regulated by authophagy. Autophagy is a ubiquitous, non-selective degradation process involving the lysosomal/vesicular pathway that protects the cells by clearing the damaged organelles and toxic protein aggregates. A fascinating suggestion which links manganese-affected homeostasis and embryo survival would involve authophagy. Future studies in this direction are needed to clarify this hypothesis.

Lastly, we described that, in echinoderm immune cells, manganese acts as a trigger for the proliferation of coelothelial cells and thus increases the number of circulating coelomocytes.

Studies on the effects of manganese on the activation of putative progenitor cells, including their proliferation and differentiation, should make an important contribution to our understanding of the operating mechanisms, and to the identification of those genes expressed before their release into the coelom.

7. Acknowledgements

This work has been supported by 60% MIUR and FSE (PON 2000/2006) grants to MCR and the BIOMINTEC Project (PITN-GA-2008-215507) to VM. The first author has been the recipient of a Doctoral fellowship from the University of Palermo. The project was the backbone of AP doctoral studies.

8. References

Aller, R.C. (1994). The sedimentary Mn cycle in long island sound: its role as intermediate oxidant and the influence of bioturbation, O2, and Corg flux on diagenetic reaction balances. *Journal of Marine Research*, Vol. 52, No. 2, (March 1994), pp. 259–295, ISSN: 1543-9542

Agnello, M., Filosto, S., Scudiero, R., Rinaldi, A.M., & Roccheri, M.C. (2006). Cadmium accumulation induces apoptosis in P. lividus embryos. *Caryologia*, Vol. 59, No. 4, pp. 403–408. Available from
http://www1.unifi.it/caryologia/past_volumes/59_4/59-4_gei11.pdf

Agnello, M., Filosto, S., Scudiero, R., Rinaldi, A.M., & Roccheri, M.C. (2007). Cadmium induces apoptotic response in sea urchin embryos. *Cell Stress & Chaperones*, Vol. 12, No. 1, (Sprinter 2007), pp. 44–50, 1355-8145

Agnello, M., & Roccheri, M.C. (2010). Apoptosis: focus on sea urchin development. *Apoptosis*, Vol. 15, No. 3, (March 2010), pp. 322-330, ISSN 1360-8185

Armstrong, N., Hardin, J., & McClay, D.R. (1993). Cell-cell interactions regulate skeleton formation in the sea urchin embryo. *Development*, Vol. 119, No. 3, (November 1993), pp. 833-840, ISSN 0950-1991

ATSDR (2008). Draft toxicological profile for manganese. Agency for toxic substances and disease registry. Division of toxicology and environmental medicine/applied toxicology branch, Atlanta, Georgia, Available from
http://www.atsdr.cdc.gov/toxprofiles/tp151-p.pdf

Au, C., Benedetto, A., & Aschner, M. (2008). Manganese transport in eukaryotes: the role of DMT1. *Neurotoxicology*, Vol. 29, No. 4, (July 2008), pp. 569–576, ISSN 0161-813X

Baden, S.P., Eriksson, S.P., & Weeks, J.M. (1995). Uptake, accumulation and regulation of manganese during experimental hypoxia and normoxia by the decapod Nephrops nor6egicus (L.). *Marine Pollution Bulletin*, Vol. 31, No. 1-3, 93-102, ISSN 0025-326X

Baden, S.P., & Neil, D.M. (1998). Accumulation of Manganese in the Haemolymph, Nerve and Muscle Tissue of Nephrops norvegicus (L.) and Its Effect on Neuromuscular Performance. *Comparative Biochemistry and Physiology*, Vol. 119A, No.1, (January 1998), pp. 351-359, ISSN 1095-6433

Baden, S.P., Eriksson, S.P., & Gerhardt, L. (1999). Accumulation and elimination kinetics of manganese from different tissues of the Norway lobster Nephrops norvegicus (L.). *Aquatic Toxicology*, Vol. 46, No. 2, (July 1999), pp. 127–137, ISSN 0166-445X

Baden, S.P., & Eriksson, S.P. (2006). Oceanography and marine biology: an annual review. In: *Role, routes and effects of manganese in crustaceans*, R.N. Gibson, R.J.A. Atkinson, J.D.M. Gordon (eds), 61-83, Taylor and Francis, ISBN 978-0-8493-7044-1, London

Balzer, W. (1982). On the distribution of iron and manganese at the sediment/water interface: thermodynamic versus kinetic control. *Geochimica et Cosmochimica Acta*, Vol. 46, No. 7, (July 1982), pp. 1153–1161, ISSN 0016-7037

Bellas, J., Fernández, N., Lorenzo, I., & Beiras, R. (2008). Integrative assessment of coastal pollution in a Ría coastal system (Galicia, NW Spain): correspondence between sediment chemistry and toxicity. *Chemosphere*, Vol. 72, No. 5, (April 2008), pp. 826–835, ISSN 0045-6535

Bonatti, E., & Nayudu, Y.R. (1965). The origin of manganese nodules on the ocean floor. *American Journal of Science*, Vol. 263, (January 1965), pp. 17-39. ISSN 1945-452X

Bonaventura, R., Poma, V., Costa, C., & Matranga, V. (2005). UVB radiation prevents skeleton growth and stimulates the expression of stress markers in sea urchin embryos. *Biochemical and Biophysical Research Communications*, Vol. 328, No. 1, (March 2005), pp. 150-157, ISSN 0006-291X

Bryan, G. W. & Ward, E. (1965). The absorption and loss of radioactive and non-radioactive manganese by the lobster, Homarus vulgaris. *Journal of the Marine Biological Association of the United Kingdom*, Vol. 45, No. 1, pp. 65-95, ISSN 0025-3154

Caccavo, F., Lonergan, D.J., Lovley, D.R., Davis, M., Stolz, J.F., & McInerney, M.J. (1994). Geobacter sulfurreducens sp-nov, a hydrogen-oxidizing and acetate-oxidizing dissimilatory metal-reducing microorganism. *Applied and Environmental Microbiology*, Vol. 60, No. 10, (October 1994), pp. 3752–3759, ISSN 0099-2240

Chiarelli, R., Agnello M., & Roccheri, M.C. (2011). Sea urchin embryos as a model system for studying autophagy induced by cadmium stress. *Autophagy*, Vol. 7, No 9, (September 2011), pp. 1028-1034, ISSN 1554-8627

CICAD (2004) Manganese and its compounds: environmental aspects. Concise international chemical assessment document 63. WHO, Geneva, Switzerland, Available from http://www.who.int/ipcs/publications/cicad/cicad63_rev_1.pdf

Coates, J.D., Lonergan, D.J., Philips, E.J.P., Jenter, H., & Lovley, D.R. (1995). Desulfuromonas palmitatis sp nov, a marine dissimilatory Fe(III) reducer that can oxidize long-chain fatty acids. *Archives of Microbiology*, Vol.164, No. 6, (December 1995), pp. 406–413, ISSN 0302-8933

Coates, J.D., Bhupathiraju, V.K., Achenbach, L.A., McInerney, M.J., & Lovley, D.R. (2001). Geobacter hydrogenophilus, Geobacter chapellei and Geobacter grbiciae, three new, strictly anaerobic, dissimilatory Fe(III)-reducers. *International Journal of Systematic and Evolutionary Microbiology*, Vol. 51, No. 2, (March 2001), pp. 581–588, ISSN 1466-5026

Colomina, M.T., Domingo, J.L., Llobet, J.M., & Corbella, J. (1996). Effect of day of exposure on the developmental toxicity of manganese in mice. *Veterinary & Human Toxicology*, Vol. 38, No. 1, pp. 7–9, ISSN 0145-6296

Couper, J. (1837). On the effects of black oxide of manganese when inhaled into the lungs. *Brain Annual Medical Pharmacology*, Vol. 1, pp. 41-42

Crowley, J.A., Traynor, D.A., & Weatherburn, D.C. (2000). Enzymes and proteins containing manganese: an overview. *Metal ions in biological systems*, Vol. 37, pp. 209-278, ISSN 0161-5149

Culotta, V. C., Yang, M., & Hall, M. D. (2005). Manganese transport and trafficking: lessons learned from Saccharomyces cerevisiae. *Eukaryotic Cell*, Vol. 4, No. 7, (July 2005), pp. 1159–1165, ISSN 1535-9778

Davidson, E.H., Cameron, R.A., & Ransick, A. (1998). Specification of cell fate in the sea urchin embryo: summary and some proposed mechanisms. *Development*, Vol.125, No. 17, (September 1998), pp. 3269-3290, ISSN 0950-1991

De Schamphelaire, L., Rabaey, K., Boon, N., Verstraete, W.,& Boeckx, P. (2007). Minireview: The potential of enhanced manganese redox cycling for sediment oxidation. *Geomicrobiology Journal*, Vol. 24, No. 7-8, 547–558, ISSN 0149-0451

Doyle, D., & Kapron, C.M. (2002). Inhibition of cell differentiation by manganese chloride in micromass cultures of mouse embryonic limb bud cells. *Toxicology in Vitro*, Vol. 16, No. 2, (April 2002), pp. 101-106, ISSN 0887-2333

Ettensohn, C.A., & Malinda, K.M. (1993). Size regulation and morphogenesis: A cellular analysis of skeletogenesis in the sea urchin embryo. *Development*, Vol. 119, No. 1, (September 1993), pp. 155–167, ISSN 0950-1991

Fernandez-Serra, M., Consales, C., Livigni, A., & Arnone, M.I. (2004). Role of the ERK mediated signaling pathway in mesenchyme formation and differentiation in the sea urchin embryo. Developmental Biology, Vol. 268, No. 2, (April 2004), pp. 384–402, ISSN 0012-1606

Gemperline, P.J., Miller, K.H., West, T.L., Weinstein, J.E., Hamilton, C.J., & Bray, J.T. (1992). Principal component analysis, trace elements, and Blue Crab Shell disease. *Analytical Chemistry*, Vol. 64, No. 9, (May 1992), pp. 523–532, ISSN 0003-2700

Geraci, F., Pinsino, A., Turturici, G., Savona, R., Giudice, G., & Sconzo, G. (2004). Nickel, lead, and cadmium induce differential cellular responses in sea urchin embryos by activating the synthesis of different HSP70s. *Biochemical and Biophysical Research Communications*, Vol. 322, No. 3, (September 2004), pp. 873–877, ISSN 0006-291X

Giordano, G., Pizzurro, D., VanDeMark, K., Guazzetti, M., & Costa, L.G. (2009). Manganese inhibits the ability of astrocytes to promote neuronal differentiation. *Toxicology and Applied Pharmacology*, Vol. 240, No. 2, (October 2009), pp. 226–235, ISSN 0041-008X

Glinski, Z., & Jarosz, J. (2000). Immune phenomena in echinoderms. *Archivum Immunologiae et Therapiae Experimentalis*, Vol. 48, No. 3, pp. 189–193, ISSN 0004-069X

Goldstone, J.V., Hamdoun, A., Cole, B.J., Howard-Ashby, M., Nebert, D.W., Scally, M., Dean, M., Epel, D., Hahn, M.E., & Stegeman, J.J. (2006). The chemical defensome: Environmental sensing and response genes in the Strongylocentrotus purpuratus genome. *Developmental Biology*, Vol. 300, No. 1, (December 2006), pp. 366-384, ISSN 0012-1606

Guss, K.A., & Ettensohn, C.A. (1997). Skeletal morphogenesis in the sea urchin embryo: regulation of primary mesenchyme gene expression and skeletal rod growth by ectoderm-derived cues. *Development*, Vol. 124, No. 10, (May 1997), pp. 1899-1908, ISSN 0950-1991

Hagiwara, S., & Takahashi, K. (1967). Surface density of calcium ion and calcium spikes in the barnacle muscle fibre membrane. *The Journal of General Physiology*, Vol. 50, No. 3, (January 1967), pp. 583–601, ISSN 0022-1295

Hamdoun, A., & Epel, D. (2007). Embryo stability and vulnerability in an always changing world. *Proceedings of the National Academy of Sciences*, Vol. 104, No. 6, (February 2007), pp. 1745–1750, ISSN 0027-8424

Hansen, S.N., & Bjerregaard, P. (1995). Manganese kinetics in the sea star Asterias rubens (L.) exposed via food or water. *Marine Pollution Bulletin*, Vol. 31, No. 1-3, pp. 127-132, ISSN 0025-326X

Hereu, B., Zabala, M., Linares, C., & Sala, E. (2005). The effects predator abundance and habitat structural complexity on survival juvenile sea urchins. *Marine Biology*, Vol. 146, No. 2, (January 2005), pp. 293-299, ISSN 0025-3162

Hernroth, B., Baden, S.P., Holm, K., André, T., & Söderhäll, I. (2004). Manganese induced immune suppression of the lobster, Nephrops norvegicus. *Aquatic Toxicology*, Vol. 70, No. 3, (December 2004), pp. 223-231, ISSN 0166-445X

Holmes, J.M., Gräns, A.S., Neil, D.M., & Baden S.P. (1999). Effects of the metal ions Mn2+ and Co2+ on muscle contraction in the Norway lobster, Nephrops norvegicus. *Journal of Comparative Physiology B*, Vol. 169, No. 6, (September 1999), pp. 402-410, ISSN 0174-1578

Holmes, D.E., Nevin, K.P., & Lovley, D.R. (2004). Comparison of 16S rRNA, nifD, recA, gyrB, rpoB and fusA genes within the family Geobacteraceae fam. nov. *International Journal of Systematic and Evolutionary Microbiology*, Vol. 54, No. 5, (September 2004), pp. 1591–1599, ISSN 1466-5026

Hunt, C.D. (1983). Variability in the benthic Mn flux in coastal marine ecosystems resulting from temperature and primary production. *Limnology and Oceanography*, Vol. 28, No. 5, pp. 913–923, ISSN 0024-3590

Krång, A.S., & Rosenqvist, G. (2006). Effects of manganese on chemically induced food search behaviour of the Norway lobster, Nephrops norvegicus (L.). *Aquatic Toxicology*, Vol. 78, No. 3, (Jun 2006), pp. 284-291, ISSN 0166-445X

Logan, C.Y., Miller, J.R., Ferkowicz, M.J., & McClay, D.R. (1999). Nuclear beta-catenin is required to specify vegetal cell fates in the sea urchin embryo. *Development*, Vol. 126, No. 2, (January 1999), pp. 345-357, ISSN 0950-1991

Matranga, V (1996). Molecular aspects of immune reactions in Echinodermata. *Progress in Molecular & Subcellular Biology*, Vol. 15, pp. 235–247, ISSN 0079-6484

Matranga, V., Pinsino, A., Celi, M., Natoli, A., Bonaventura, R., Schröder, H.C., & Müller, W.E.G. (2005). Monitoring chemical and physical stress using sea urchin immune cells. *Progress in Molecular & Subcellular Biology*, Vol. 39, pp. 85–110, ISSN 0079-6484

Matranga, V., Zito, F., Costa, C., Bonaventura, R., Giarrusso, S., & Celi, F. (2010). Embryonic development and skeletogenic gene expression affected by X-rays in the Mediterranean sea urchin Paracentrotus lividus. *Ecotoxicology*, Vol. 19, No. 3, (March 2010), pp. 530–537, ISSN 0963-9292

Matranga, V., Bonaventura, R., Costa, C., Karakostis, K., Pinsino, A., Russo, R., & Zito, F. (2011). Echinoderms as blueprints for biocalcification: regulation of skeletogenic genes and matrices. *Progress in Molecular and Subcellular Biology*, Vol. 52, pp. 225-248, ISSN 0079-6484

Mena, I., Mario, O., Fuenzalida, S., & Cotzias, G.C. (1967) Chronic manganese poisoning-clinical picture and manganese turnover. *Neurology*, Vol. 17, No. 2, (February 1967), pp. 128-136, ISSN 0028-3878

Middelburg, J.J., & Levin, L.A. (2009). Coastal hypoxia and sediment biogeochemistry. *Biogeosciences*, 6, 3655–3706. Available from www.biogeosciences-discuss.net/6/3655/2009/

Missiaen, L., Raeymaekers, L., Dode, L., Vanoevelen, J., Baelen, K.V., Parys, J.B., Callewaert, G., Smedt, H.D., Segaert, S., & Wuytack, F. (2004). SPCA1 pumps and Hailey-Hailey disease. *Biochemical and Biophysical Research Communications*, Vol. 322, No. 4, (October 2004), pp. 1204–1213, ISSN 0006-291X

Muñoz-Chápuli, R., Carmona, R., Guadix, J.A., Macías, D., & Pérez-Pomares, J.M. (2005). The origin of the endothelial cells: an evo-devo approach for the invertebrate/vertebrate transition of the circulatory system. *Evolution & Development*, Vol. 7, No. 4, (July-August 2005), pp. 351–358, ISSN 1520-541X

Normandin, L., & Hazell, A.S. (2002). Manganese neurotoxicity: An update of pathophysiologic mechanisms. *Metabolic Brain Disease*, Vol. 17, No. 4, (December 2002), pp. 375–387, ISSN 0885-7490

Nørum, U., Lai, V.W., & Cullen, W.R. (2005). Trace element distribution during the reproductive cycle of female and male spiny and Pacific scallops, with implications for biomonitoring. *Marine Pollution Bulletin*, Vol. 50, No. 2, (February 2005), pp. 175–184, ISSN 0025-326X

Oweson, C.A., Baden, S.P., & Hernroth, B.E. (2006). Manganese induced apoptosis in haematopoietic cells of Nephrops norvegicus (L.). *Aquatic Toxicology*, Vol. 77, No. 3, (May 2006), pp. 322-328, ISSN 0166-445X

Oweson, C., Sköld, H., Pinsino, A., Matranga, V., & Hernroth, B. (2008). Manganese effects on haematopoietic cells and circulating coelomocytes of Asterias rubens (Linnaeus). *Aquatic Toxicology*, Vol. 89, No. 2, (August 2008), pp. 75-81, ISSN 0166-445X

Oweson, C., & Hernroth B. (2009). A comparative study on the influence of manganese on the bactericidal response of marine invertebrates. *Fish & Shellfish Immunology*, Vol. 27, No. 3, (September 2009), pp. 500-507, ISSN 1050-4648

Oweson, C., Li, C., Söderhäll, I., & Hernroth, B. (2010). Effects of manganese and hypoxia on coelomocyte renewal in the echinoderm, Asterias rubens (L.). *Aquatic Toxicology*, Vol. 100, No. 1, (October 2010), pp. 84-90, ISSN 0166-445X

Perl, D.P., & Olanow, C.W. (2007). The neuropathology of manganese-induced parkinsonism. *Journal of Neuropathology & Experimental Neurology*, Vol. 66, No. 8, (August 2007), pp. 675–682, ISSN 0022-3069

Pinsino, A., Thorndyke, M.C., & Matranga, V. (2007). Coelomocytes and post-traumatic response in the common sea star Asterias rubens. *Cell Stress & Chaperones*, Vol. 12, No.4, (Winter 2007), pp. 331–341, ISSN 1355-8145

Pinsino, A., Matranga, V., Trinchella, F., & Roccheri, M. C. (2010). Sea urchin embryos as an in vivo model for the assessment of manganese toxicity: developmental and stress response effects. *Ecotoxicology*, Vol. 19, Vo. 3, (March 2010), pp. 555-562, ISSN 0963-9292

Pinsino, A., Turturici, G., Sconzo, G., & Geraci, F. (2011). Rapid changes in heat-shock cognate 70 levels, heat-shock cognate phosphorylation state, heat-shock transcription factor, and metal transcription factor activity levels in response to heavy metal exposure during sea urchin embryonic development. *Ecotoxicology*, Vol. 20, No.1, (January 2011), pp. 246-254, ISSN 0963-9292

Pinsino, A., Roccheri, M. C., Costa, C., & Matranga, V., (2011). Manganese interferes with calcium, perturbs ERK signalling and produces embryos with no skeleton. *Toxicological Sciences*, Vol. 123, No. 1, (September 2011), pp. 217-30, ISSN 1096-6080

Regoli, F., Orlando, E., Mauri, M., Nigro, M., & Cognetti, G.A. (1991). Heavy metal accumulation and calcium content in the bivalve Donacilla cornea. *Marine Ecology Progress Series*, Vol. 74, No 2-3, (August 1991), pp. 219–224, ISSN 0171-8630

Roccheri, M.C., Agnello, M., Bonaventura, R., & Matranga, V. (2004). Cadmium induces the expression of specific stress proteins in sea urchin embryos. *Biochemical and Biophysical Research Communications*, Vol. 321, No. 1, (August 2004), pp. 80-87, 0006-291X

Roccheri, M.C., & Matranga, V. (2010). Cellular, Biochemical and molecular effects of cadmium on marine invertebrates: focus on Paracentrotus lividus sea urchin development. In: Cadmium in the Environment, R.G. Parvau (ed), 337-366, Nova Science Publishers, Inc., ISBN: 978-1-60741-934-1, United States of America

Rosen, G., Rivera-Duarte, I., Chadwick, D.B., Ryan, A., Santore, R.C., & Paquin, P.R. (2008). Critical tissue copper residues for marine bivalve (Mytilus galloprovincialis) and echinoderm (Strongylocentrotus purpuratus) embryonic development: conceptual, regulatory and environmental implications. *Marine Environmental Research*, Vol. 66, No. 3, (September 2008), pp. 327–336, ISSN 0141-1136

Roth, J.A. (2006). Homeostatic and toxic mechanisms regulating manganese uptake, retention, and elimination. *Biological Research*, Vol. 39, No. 1, (April 2005), pp. 45–57, ISSN 0716-9760

Röttinger, E., Besnardeau, L., & Lepage, T. (2004). A Raf/MEK/ERK signalling pathway is required for development of the sea urchin embryo micromere lineage through phosphorylation of the transcription factor Ets. Development, Vol. 131, No. 5, (March 2004), pp. 1075-1087, ISSN 0950-1991

Russo, R., Bonaventura, R., Zito, F., Schroder, H.C., Muller, I., Muller, W. E.G., & Matranga, V. (2003). Stress to cadmium monitored by metallothionein gene induction in Paracentrotus lividus embryos. *Cell Stress & Chaperones*, Vol. 8, No. 3, pp. 232–241, ISSN 1355-8145

Russo, R., Zito, F., Costa, C., Bonaventura, R., & Matranga, V. (2010). Transcriptional increase and misexpression of 14-3-3 epsilon in sea urchin embryos exposed to UV-B. *Cell Stress & Chaperones*, Vol. 15, No. 6, (November 2010), pp. 993-1001, ISSN 1355-8145

Saby, E., Justesen, J., Kelve, M., & Uriz, M.J. (2009). In vitro effects of metal pollution on Mediterranean sponges: species-specific inhibition of 2',5'-oligoadenylate synthetase. *Aquatic Toxicology*, Vol. 94, No. 3, (September 2009), pp. 204-210, ISSN 0166-445X

Sanchez, D.J., Domingo, J.L., Llobet, J.M., & Keen, C.L. (1993). Maternal and developmental toxicity of manganese in the mouse. *Toxicology Letters*, Vol. 69, No. 1, (July 1993), pp. 45-52, ISSN 0378-4274

Santamaria, A.B. (2008). Manganese exposure, essentiality and toxicity. *Indian Journal of Medical Research*, Vol. 128, No. 4, (October 2008), pp.484–500, ISSN 0971-5916

Schamphelaire, L., Rabaey, K., Boeckx, P.,Boon, N., & Verstrate, W. (2008). Outlook for benefits of sediment microbial fuel cells with two bio-electrodes. *Microbial Biotechnology*, Vol. 1, No. 6, (November 2008), pp. 446-462, ISSN 1751-7907

Sea Urchin Genome Sequencing Consortium (2006). The genome of the sea urchin Strongylocentrotus purpuratus. *Science*, Vol. 314, No. 5801, (November 2006), pp. 941–952, ISSN 1095-9203

Smith, L., Ghosh, J., Buckley, K.M., Clow, L.A., Dheilly, N.M., Haug, T., Henson, J.H., Li, C., Lun, C.M., Majeske, A.J., Matranga, V., Nair, S.V., Rast, J.P., Raftos, D.A., Roth, M., Sacchi, S., Schrankel, C.C., & Stensvåg, K. (2010). Echinoderm Immunity. In: *Invertebrate Immunology*, Soderhall K (ed), pp. 261-301, Landes Bioscience, Available from
http://www.gwu.edu/~biology/faculty/SmithPapers/Smith%20etal%20Echinod ermImmunityChapter%202010.pdf

Sunda, W.G., & Huntsman, S.A. (1998). Processes regulating cellular metal accumulation and physiological effects: Phytoplankton as model systems. *Science of the Total Environment*, Vol. 219, No. 2-3, (August 1998), pp. 165–181, ISSN 0048-9697

Takeda, A. (2003). Manganese action in brain function. *Brain Research Reviews*, Vol. 41, No. 1, (Jenuary 2003), pp. 79-87, ISSN 0165-0173

Trefry, J.H., Presley, B.J., Keeney-Kennicutt, W.L., & Trocine, R.P. (1984). Distribution and chemistry of manganese, iron, and suspended particulates in orca basin. *Geo-Marine Letters*, Vol. 4, No. 2, pp. 125–130, ISSN 0276-0460

Vandieken, V., Mussmann, M., Niemann, H., & Jorgensen, B.B. (2006). Desulfuromonas svalbardensis sp nov and Desulfuromusa ferrireducens sp nov., psychrophilic, Fe(III)-reducing bacteria isolated from Arctic sediments, Svalbard. *International Journal of Systematic and Evolutionary Microbiology*, Vol. 56, No. 5, (May 2006), pp. 1133–1139, ISSN 1466-5026

Wang, X., Gan, L., Wiens, M., Schloßmacher, U., Schröder, H.C., & Müller, W.E. (2011). Distribution of Microfossils Within Polymetallic Nodules: Biogenic Clusters Within Manganese Layers. *Marine Biotechnology*, (May 31), [Epub ahead of print], ISSN 1436-2236

Wedler, F. (1994). Biochemical and nutritional role of manganese: an overview. In: *Manganese in health and disease*, D. Klimis-Tavantzis (ed), CRC Press, 1–38, ISBN: 9780849378416, Boca Raton, Florida

Wehrli, B., Friedl, G., & Manceau, A. (1995) Reaction Rates and Products of Manganese Oxidation at the Sediment-Water Interface, In: *Aquatic Chemistry: Principles and Applications of Interfacial and Inter-Species Interactions in Aquatic Systems*, C.P. Huang, C. O' Melia, J.J. Morgan (eds), Washington, 111-134, ISBN: 9780841224261, American Chemical Society

Weinstein, J.E., West, T.L., & Bray, J.T. (1992). Shell disease and metal content of blue crabs, Callinectes sapidus, from the Albemarle-Pamlico estuarine system, North Carolina. *Archives of Environmental Contamination and Toxicology*, Vol. 23, No. 3, (October 1992), pp. 355-362, ISSN 0090-4341

Zito, F., Tesoro, V., McClay, D.R., Nakano, E., & Matranga, V. (1998). Ectoderm cell--ECM interaction is essential for sea urchin embryo skeletogenesis. *Developmental Biology*, Vol. 96, No. 2, (April 1998), pp. 184-192, ISSN 0012-1606

Zito, F., Costa, C., Sciarrino, S., Cavalcante, C., Poma, V., & Matranga, V. (2005). Cell adhesion and communication: a lesson from echinoderm embryos for the exploitation of new therapeutic tools. *Progress in Molecular & Subcellular Biology*, Vol. 39, pp. 7–44, ISSN 0079-6484

Response of C_3 and C_4 Plant Systems Exposed to Heavy Metals for Phytoextraction at Elevated Atmospheric CO_2 and at Elevated Temperature

Jatin Srivastava[1], Harish Chandra[2],
Anant R. Nautiyal[3] and Swinder J. S. Kalra[4]
[1]*Department of Applied Sciences, Global Group of Institutions*
Raebareli Road Lucknow UP
[2]*Department of Biotechnology, G. B. Pant Engineering College*
Ghurdauri, Pauri Garhwal, Uttrakhand
[3]*High Altitude Plant Physiology Research Centre*
H. N. B. Garhwal University, Srinagar Garhwal, Uttrakhand
[4]*Department of Chemistry, Dayanand Anglo Vedic College, Civil-Lines, Kanpur UP*
India

1. Introduction

Increasing concentration of CO_2 in the lower atmosphere and an increase in average annual temperature are two major factors associated with global climate change. Researches show profound impact of climate change on the global primary productivity (Krupa, 1996; Kimball, 1983). Apart from the climate change, environmental contamination especially of those chemicals are non degradable and persist in the environment for long e.g., chlorinated pesticides, and heavy metals. In fact, heavy metal ions such as Cu^{2+}, Zn^{2+}, Mn^{2+}, Ni^{2+}, Fe^{2+}, and Co^{2+} are essential in trace amounts for growth of organisms (Kunze et al., 2001; Choudhry et al., 2006), however; the excessive amounts of these metal ions along with other non essential metals such as Pb^{2+}, Cd^{2+}, and Hg^{2+} aggravated the menace to an alarming stage (Kamal et al., 2004). Our knowledge regarding the non degradable contaminants has enhanced in last two decades however; more researches are needed for the mitigation and to reduce the introduction of any new contaminant in to the environment. Researchers all over the world endeavoured to resolve the problem and have developed several techniques to restore the quality of environment. The only solution to the problem associated with non degradable contaminants is the removal from the contaminated sites (Lasat, 2002).

Phyto-remediation has achieved a top priority among the activists and scientists because of its cost effectivity and sustainable nature. Voluminous literature based on extraneous researches is available today supporting the use of plant systems for the removal of heavy metals from the contaminated area (Khan et al., 1998; Ebbs & Kochian, 1998; Hinchman et al., 1998; Srivastava & Purnima, 1998; Pulford & Watson, 2003; Wilde et al., 2005) and the references quoted there in. Since plants respond differently in altered environmental conditions viz., nutrient status of soil, water availability, pollution, elevated CO_2 and elevated temperature and humidity, it is imperative to review the performances and responses of plant systems especially when growing in contaminated sites with heavy metals at extreme environmental conditions like that are posed by the global climate change. However; prior to review the responses of C_3 and C_4 plant systems lets consider the general aspects regarding heavy metals, contamination, toxicity and phytoremediation.

1.1 Heavy metals (HMs): Contamination and toxicity

Heavy metals (HMs) form the main group of inorganic contaminants (Alloway, 1990). There are number of instances worldwide polluted with HMs naturally as well as anthropogenically. Industries especially metallurgical, foundries, mining, tanneries and thermal power plants have significant importance as these generate huge amounts of waste containing higher concentrations of toxic metals (Gupta et al., 2010; Aguilera et al., 2010). HMs are defined as the transition elements having density more than 5 gm cm^{-3}, and recently have been defined as the Class – B (border line) (Nieboer & Richardson, 1990). Out of 90 naturally occurring elements, 53 are heavy metals (Weast, 1984) out of which only few (around 17) metals are biologically significant because these are readily available to living cells (Weast, 1984; Pickering, 1995). Action of HMs leading toxicity in living cells depending upon several physico-chemical factors such as redox potential (of surrounding medium and inside the cells). Because of being transitional elements, several ions of different valence states are quite common of same metal for e.g., four species of chromium (Cr) viz., Cr, Cr^{3+}, Cr^{4+}, and Cr^{6+}. Most of the HMs usually form cations e.g., Cu^{2+}, Zn^{2+}, Mn^{2+}, Ni^{2+}, Fe^{2+}, and Co^{2+} (table 1). There exist two groups of HMs viz., redox active and redox inactive (Schützendübel & Polle, 2002). Once toxic metals are present in the environment they eventually become a part of abiotic and biotic components of an ecosystem (Galloway et al., 1982), posing toxicity to the living organisms interacting with each other. Metals with low redox potential (Eh value) has very little significance for biological redox reactions. Auto-oxidation and Fenton type reactions are supposed to be the cause of free radicals generation especially the reactive oxygen species (ROS) from HMs causing injury the cells and cell organelles (Jones et al., 1991; Shi et al., 1993; Stohs & Bagchi, 1995). Heavy metals are especially toxic because of their ability to bind with proteins and prevent DNA replication as these can bind strongly to oxygen, nitrogen and sulphur atoms (Nieboer & Richardson, 1980) and inactivate enzymes by binding to cysteine residues (Schützendübel & Polle, 2002).

Common occurring metals	Toxic ionic species
Cd (Cadmium)	Cd^{2+}
Co (Cobalt)	Co^{2+}
Cu (Copper)	Cu^{2+}, Cu^{3+}
Cr (Chromium)	Cr^{6+}
Fe (Iron)	Fe^{3+}
Hg (Mercury)	Hg^{2+}, Hg^{3+}
Mn (Manganese)	Mn^{2+}, Mn^{3+}
Ni (Nickel)	Ni^{2+}, Ni^{3+}
Pb (Lead)	Pb^{2+}
Zn (Zinc)	Zn^{2+}

Table 1. Heavy metals of common occurrence and their toxic ionic species

1.2 Phyto-remediation of HMs: General aspects

Removal of any non-degradable, undesirable, inorganic or organic contaminant or pollutant with the help of plants is commonly termed as phytoextraction and the process is called phyto-remediation. Phyto-remediation is a cost effective and a sustainable way to mitigate the environmental pollution that has attracted scientists and policy makers all over the world. The only remedy for heavy metal contamination is to remove and reuse (Chojnacka, 2010). Phytoextraction, rhizo-filtration and phytostabilization are the technical processes occur in a plant simultaneously to make up *in-situ* phyto-remediation (Suresh & Ravishankar, 2004). Phyto-remediation technique has technical advancement over the traditional chemical and physical techniques to remove contaminants from the environment (Garbisu & Alkorta, 2001). Phyto-mining which signifies the recovery of rare and valuable trace metal contaminants from the harvested biomass (net primary product) offers a great significance in metal removal. Metal contamination is removed by plants capable of defending the toxic manifestations by three distinct ways viz., (1) restrict entry of metals in to the soft growing tissues by excluding metals from the metabolic pathways, (2) restrict entry to the shoot as the metals is accumulated by the roots, (3) accumulation of metals in different parts (Kamal et al., 2004). Successful restoration is however; dependent on the selection of plant species based on the method of their establishment along with the knowledge of growth regulating factors (Tu & Ma, 2003). Ideally plants, growing fast, capable of producing higher biomass, and able to tolerate and accumulate high concentrations of HMs in shoots are best suited for phyto-remediation. *Brassicaceae* family of C₃ plants have metal accumulating capability to greater extent (Kumar et al., 1995). Phyto-remediation is most useful when contaminants are within the root zone of the plants i.e., top soil (up to 1 meters) (Wilde et al., 2005). Biochemically plants are equipped to protect their selves from the toxicity of metals as they synthesize Cys-rich (Cysteine rich), metal – binding peptides including phyto-chelatins and metallothioneins (-SH group containing peptides) (Jonak et al., 2004) to relocate HMs by chelation and sequestration in the vacuole (Clemens, 2001) on the other hand, membrane transport systems provide plants tolerance for toxic metals (Hall & Williams, 2003).

2. C₃ and C₄ plant system: An introduction

In nature three different plant systems exist viz., C₃, C₄ and CAM, characterized on the basis of CO₂ trapping mechanisms, however; C₄ and CAM plants essentially follow C₃ pathway to trap CO₂ as an initial step. C₄ is a characteristic photosynthesis syndrome of angiosperms. In general *phosphoenolpyruvate carboxylase* (EC 4.1.1.31, PEPC) enzyme is widespread among all plants, including C₃ (e.g., *Pisum sativum, Gossipium hirsutum, Oryza sativa, Brassica campestris, Triticum aestivum, Avena sativa*) C₄ (e.g., *Zea mays, Saccharum officinarum, Sorghum spp., Vetiveria zizanioides, Cyanadon dactylon*) and CAM (e.g., members of Orchidaceae, Polypodiaceae (ferns)) species and is responsible for the initial carbon fixation in C₄ and CAM plants (O' Leary, 1982). CAM plants are very few in nature and have very little significance for this review purpose. C₃ and C₄ plants have unique carbon trapping mechanisms. The general enzymatic system involves in CO₂ fixation in C₃ and C₄ are Ribulose-1-5-bisphosphate carboxylase oxygenase (Rubisco EC 4.1.1.39) and phosphoenolpyruvate carboxylase (PEPC), NADP-malic enzyme (NADP-ME), Pyruvate, phosphate dikinase (PPDK) respectively. The leaves of C₄ plants display Kranz anatomy whereby an outer layer of mesophyll cells containing chloroplast surrounds vascular bundles with an inner layer of bundle sheath cells (Dengler & Nelson, 1999). In C_3 plants mesophyll cells are devoid of chloroplast and CO_2 is fixed in bundle sheath cells by Rubisco. Chloroplasts of C_3 plant contain a complete Calvin cycle and are able to assimilate CO_2 to convert it to the principle 3 carbon compound (triose phosphate), on the other hand CO_2 is distributed in two cells viz., mesophyll and bundle sheath in C_4 plants and converted primarily in 4 carbon compound (acid oxaloacetate) by the action of PEP-C in mesophyll cells (Ueno, 2001) which is then transported in bundle sheath cells where by the acids from mesophyll cells provide carbon dioxide. Extensive research literature is available on the comparative account of C_3 and C_4 plant systems (Du & Fang, 1982; Rajendrudu & Das, 1982; Matsuoko & Hata, 1987; Wand et al., 1999; Ueno, 2001; Winslow et al., 2003; Derner et al., 2003; Sage, 2004; Edwards et al., 2005; Niu et al., 2006; Caird et al., 2007; Bräutigam et al., 2008; Tang et al., 2009; Weber & Caemmerer, 2010, Doubnerová, & Ryšlavă, 2011) however; very few and scattered information is available on the comparative account of C_3 and C_4 plants and their growth performances under extreme environmental conditions. In this chapter, collective information based on the established researched facts from all over the world regarding the responses of C_3 and C_4 plants growing under stressful environment have been reviewed.

C_4 plants have higher rates of photosynthesis than C_3 plants (Sage, 2004; Weber & Caemmerer, 2010). Photosynthesis in C_4 plants does not saturate but increases at high light intensities and can continue at very low CO_2 concentrations (Sage, 2004; Bräutigam et al., 2008). Subsequently, these plants have rapid growth rates and higher biomass and economic yields as compared to the C_3 plants. There are evidences from researches that C_4 plant such as vetiver grass (*Vetiveria zizanioides* L. Nash) can withstand harsh environmental conditions (Truong & Baker, 1998, Chen et al., 2004, Chiu et al., 2006; Srivastava et al., 2008). A comparative study performed on two separate species belonging to C_3 and C_4 systems respectively show that the environmental tolerance depends on the high biomass production which is higher in case of C_4 plants (Ye et al., 1997; Ali et al., 2002). However; there is lack of information regarding the biochemical differences among C_3 and C_4 plant systems exposed to toxic environment for e.g., the extent of detoxification mechanism, mycorrhization, proteomes (expression of genes). The researches carried out for the investigations of toxic response of particular plant variety belonging to C_3 and C_4 type indicate that there is a high tolerance in C_4 plants as compared to the C_3 plants which may or may not be true for the entire group of plants belonging to these systems (Chapin, 1991; Ali et al., 2002; Niu et al., 2006). C_4 photosynthesis allows fast biomass accumulation with high nitrogen and water use efficiency (Leegood & Edwards, 1996; Sage, 2004) which is desired set of traits to increase the productivity of crop plants (Matsuoka et al., 1998) and a required character for successful phytoremediation.

3. Plant's response, environmental contamination and environmental factors

Prior to study the response of plant systems for any particular measurable environmental factor such as heavy metal concentration in water or soil, or set such as soil related factors including physical as well as chemical, it is required to take measurable responsive quantity such as biomass, growth rate, and enzyme kinetics. Small changes in environmental conditions may cause significant alterations in growth rates therefore growth rate is preferred as the primary and essential parameters to monitor impact of any environmental factor. In general plants achieve variety of mechanisms providing tolerance or resistance against environmental heights. The locally adapted plants (Eco-types) are well versed with such mechanisms making the species capable to survive in their corresponding environmental conditions. The response of plants (C_3 and C_4 plant systems) exposed to any undefined environmental extremity should not be considered alone because the response of a plant against any (favorable or stressful) conditions is a result of collective influence of stress and prevailed environmental conditions for e.g., in warmer conditions C_3 plants do not respond positively however; C_4 does and vice versa which is evident from the studies of C_3 and C_4 intermediate plant species such as *Phragmites australis* (Zheng et al., 2000) and *Eleocharis vivipara* (Ueno, 2001). Continuous altering weather conditions, soil related factors, light intensity, water availability, nutrient status, temperature, humidity, and evaporation coefficient make the *in-situ* experimentation very difficult and the response measurement of plants becomes dubious and assignment of the reason for a particular response becomes tedious affair. There are many other factors that may or may not be responsible for a particular response, for e.g., plants growing in mine tailings may exhibit negative growth impacts, a response which is usually attributed to the presence of toxic metals however; the influence of prevailed environmental conditions remain silent which must be addressed. In general environmental stress conditions alter the plant metabolism e.g., photosynthesis.

4. Affect of climate change on C_3 and C_4 plant systems

Studies show that elevated CO_2 often enhances biomass more in C_3 (41 – 44%) than in C_4 plants (22 – 33%) (Poorter, 1993; Wand et al., 1999) however; more advanced studies suggest that certain environmental factors such as water and nutrient availability can modify plant response to CO_2 enrichment (Oren et al., 2001; Derner et al., 2003; Hikosaka et al., 2005). Although increased biomass in response to elevated CO_2 are often greater in C_3 plants, C_4 plants are often more resistant to high temperature and better adapt to low nutrient environments (Edwards et al., 2005). This suggests that even with elevated CO_2 levels, C_4 plants are able to maintain their competitive advantage over C_3 plants under environmental extremities (table 2). No any direct evidence is available on the responses of C_3 and C_4 plants used for environmental mitigation purposes at elevated CO_2 and temperature. However; the researches carried out for the investigations of toxic response of plants belonging to C_3 and C_4 system indicate high tolerance in C_4 plants as compared to the C_3 plants that can not be stipulated as a generalization for the entire group of plants belonging to these systems (Leegood and Edwards, 1996). C_4 photosynthesis allows fast biomass accumulation with high nitrogen and water use efficiency (Leegood and Edwards, 1996; Sage, 2004) and is a desired trait to increase the productivity of crop plants (Matsuoka et al., 1998). Studies of Ehleringer & Björkman (1977) showed that C_3 plants were favored at low temperature while the high temperature favored C_4 plants. The response of C_3 and C_4 plants and their global distribution has been proved to be a function of temperature (Lloyd & Farquhar, 1994). However; extensive research reports are available on the temperature and distribution of the C_3 and C_4 plants (Dickinson & Dodd, 1976; Ehleringer et al., 1997; Sage et al., 1999; Winslow et al., 2003).

In general C_3 plants are favored by low temperature thus distributed more at higher altitudes and C_4 plants are favored by high temperature. Riesterer et al. (2000) suggested that essentiality of water availability for both (C_3 and C_4) plant systems to grow under seasonal temperature variations. In C_4 plant system photo-respiration is rarely greater than 5% of the rate of photosynthesis; in C_3 plants, it can exceed 30% of the rate of photosynthesis above 30ºC (Sage & Pearcy, 2000). Because of such improvements, C_4 plants have higher productive potential, and greater light, water and nitrogen use efficiency of both photosynthesis and growth (Evans, 1993; Brown, 1999).

Response Parameters	C_3 under elevated CO_2 & temperature	C_4 under elevated CO_2 & temperature	References
Photosynthetic activity	Cold conditions preferred	Higher temperature resistant	Weber & Caemmerer, 2010; Ueno, 2001
Photorespiration	Can exceed 30%	Hardly achieve 5%	Sage, 2004
Light use efficiency	Lesser	Greater	Bräutigam et al., 2008; Evans, 1993
Biomass (gm dry wt.)	Slightly < C_4 (33%)	Slightly > C_3 (44%)	Sage, 2004; Wand et al., 1999
Mycorrhization	Lesser	Higher	Tang et al., 2009; Treseder, 2004
Water use efficiency	Less efficient	Highly efficient	Derner et al., 2003; Winslow et al., 2003
Nitrogen Use efficiency	Less efficient	Highly efficient	Niu et al., 2006; Edwards et al., 2005
Stomatal conductance	High	Lower	Caird et al., 2007

Table 2. Performance of C_3 and C_4 plants under elevated carbon dioxide (CO_2) and at elevated temperature

5. Climate change and microbial association in C_3 and C_4 plant systems

Arbuscular mycorrhizal (AM) association with roots of plants and is one of the well accepted factors responsible for their growth on disturbed sites such as heavy metal contaminated soils (Khade & Adholeya, 2009). In addition, microbial associations often provide some sort of immunity (Srivastava et al., 2008) to plants against the environmental extremities (Gaur & Adholeya, 2004). A unique feature added to plants is that, elevated CO_2 often increases in mycorrhizal colonization in the roots (Rilling et al., 1998; Treseder, 2004) because elevated CO_2 enhances carbon allocation to roots (Rilling & Allen, 1998). Studies also showed increase in mycorrhizal infection rate with elevated CO_2 levels (Staddon & Fitler, 1998; Treseder, 2004; Hu et al., 2005). Monz et al. (1994) reported that the enhancement of mycorrhizal colonization by elevated CO_2 was higher in C_4 plants than in C_3 plants. This CO_2 enhanced mycorrhizal colonization may alter plant nutrient uptake and plant interactions with their neighbors (O'Conner et al., 2002; Chen et al., 2007) particularly if mycorrhizae in the coexisting species respond differently to CO_2. Interestingly, C_4 plants are more dependent on mycorrhizae for growth and productivity than C_3 plants (Wilson & Hartnett, 1998). Altered responses have been reported in C_3 and C_4 plants because of mycorrhizae under low nutrient environment (Tang et al., 2006). In agricultural systems most troublesome weeds especially those are difficult to control belong to C_4 plant systems,

Response of C_3 and C_4 Plant Systems Exposed to Heavy Metals for Phytoextraction at Elevated Atmospheric CO_2 and at Elevated Temperature

29

while most of the major crops are C_3 plants (Patterson, 1995; Fubrer, 2003). Tang et al. (2009) reported the CO_2 enhanced arbuscular mycorrhiza (AM) affect competition for light between C_4 and C_3 plants resulting in that shoot biomass of C_4 plants was higher than in C_3 under elevated CO_2.

6. Biochemical response evaluation of C_3 and C_4 plant systems exposed to heavy metal stress under elevated CO_2 and temperature

In general, the growth of plants increase under elevated atmospheric CO_2 (Poorter et al., 1996), depending upon the prevailing trends of certain anionic nutrients such as PO_4^{-3}, and NO_3^{-1}. Voluminous literature in this regard indicates that plants growing well even under stressful conditions if provided with nutrients. Since C_4 plants have biochemical advantages over C_3 plant systems (table 3) as C_4 photosynthesis is characterized by excessive CO_2 at the site of Rubisco, that helps reducing the rate of photorespiration and increasing net carbon

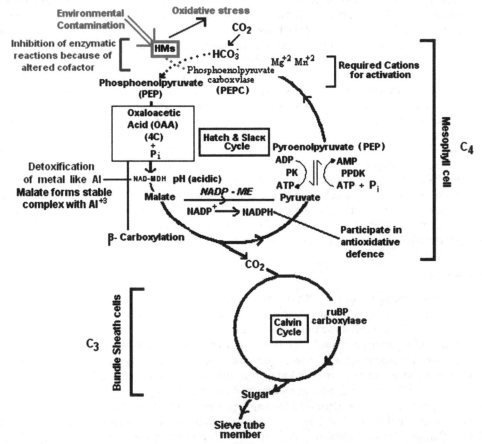

Fig. 1. Photosynthesis diagram of C_3 and C_4 plant showing the role of PEPC enzyme in heavy metal response whereas Phosphoenolpyruvate carboxylase (PEPC) enzyme acts as metal detoxifying agent.

dioxide assimilation, in other words the biomass. The plants viz., C_3 and C_4 exposed to HMs stress under elevated atmospheric CO_2 and at elevated temperature follow the same trend as in normal conditions however; there are significant alterations in biochemistry of photosynthesis of both types of plant systems. Figure 1 shows the effect of HMs on the enzymes those catalyses the photosynthetic reactions e.g., PEPC, PPDK (pyruvate phosphoenol dikinase), NADP-ME (NADP dependent malic enzyme) (Doubnerová, & Ryšlavá, 2011). PEPC enzyme that catalyze the reaction of bicarbonate and phosphoenol pyruvate (PEP) to yield 4 C acid compound oxaloacetic acid (OAA) need divalent metal ions such as Mg^{+2} and Mn^{+2} for activation. These metal ions along with other such as Fe and Zn are cofactors and may be replaced by the HMs that inhibit the activity of enzyme resulting into less CO_2 supply at the rubisco site thereby reducing the net productivity.

Elevated atmospheric CO_2 can disrupt the pH (alkalization) of cell sap by forming HCO_3^{-1}, which induces the activity of PEPC. Despite having toxic manifestations of HMs, the abundant PEPC favors the production of OAA in C_4 plants, which is further converted into acidic malate by the action of NAD dependent malate dehydrogenase enzyme (NAD-MDH EC 1.1.1.37). This NAD-MDH enzyme also act as detoxifying agent as it also catalyze the formation of stable compounds of malate and metals such as aluminum (Ma & Furukawa, 2003). C_3 plants however; are devoid of such defense mechanisms as PEPC is not the primary carbon dioxide fixing enzymes and face the oxidative stress caused by HMs. Since photosynthetic efficiency depends largely on the activity of rubisco (Sage, 2004), elevated atmospheric CO_2 increases the photorespiration in C_3 plants as a result of oxygenase activity of rubisco increasing the oxidative conversion of metals present in the cell. The oxidative conversion of metals in C_3 plant is also favored by the higher atmospheric temperature (Du & Fang, 1982). The oxidation of metals in living organisms is thiol (-SH) containing compound mediated with the hydrogen peroxide (H_2O_2) catalase complex. Catalase is present in peroxisomes of plant cells. In C_3 plants high photorespiration rates as a result of elevated concentration of CO_2, peroxisomes are large and numerous while in the C_4 plants the peroxisomes of mesophyll cells are small and fewer in number (Tolbert, 1971). The *milieu interior* of cells of C_4 plants is highly reductive thus prevent the oxidative conversion of metals.

7. Conclusion

Global climate change coupled with the environmental contamination with non-degradable substances such as pesticides, organic compounds and heavy metals are the well known fact of today's world. Among these the heavy metal contamination is inevitable in the modern age because of rapid urbanization. The health hazards of this non-degradable environmental contaminant are alarming as evident from the researches all over the world. The only solution is to extract or remove metal ions from the contaminated media. Phytoextraction is the most cost effective and environmentally sound technology for the removal of heavy metals whereby plants are employed to remove safely the toxic metal ions. C_3 and C_4 both are well researched for their ability to survive the environmental extremities such as that of climate change. Heavy metals are taken up by both types of plants however; we conclude with the findings that C_4 are the most appropriate plant system for phytoextraction although there are few benefits associated with C_3 such as higher biomass under elevated atmospheric carbon dioxide, which is required characteristic for removal of environmental contaminants. In addition to this microbial association is also favored at elevated CO_2 that provide tolerance to the plants therefore only biomass can not be ascertained as the measure responsible for the survival of C_3 or C_4 plants in conditions like global climate change.

8. References

Aguilera, I. Daponte, A. Gil, F. Hernández, A.F. Godoy, P. Pla, A. Ramos, J.L. (on behalf of the DASAHU group) (2010). Urinary levels of arsenic and heavy metals in children and adolescents living in the industrialised area of Ria of Huelva (SW Spain). *Environment International* 36: 563 – 569. ISSN No. 0160 – 4120

Ali, N.A. Bernal, M.P. Ater, M. (2002). Tolerance and bioaccumulation of copper in *Phragmites australis* and *Zea mays*. *Plant and Soil* 239: 103 – 111. ISSN No. 1573 – 5036

Alloway, B.J. (1990) Soil processes and behaviour of metals. In: Alloway, B.J. (Ed.) *Heavy metals in soils* Second ed. Blackie Academic and Professional, Chapman and Hall, Glasgow. ISBN No. 0-7514-0198-6

Bräutigam, A. Hoffmann-Benning, S. Comparative proteomics of choloroplast envelops from C_3 and C_4 plants reveals specific adaptations of plastid envelope to C_4 photosynthesis and candidate protein *Plant Physiology* 148: 568 – 579 ISSN No. 1532 – 2548

Brown, R.H. (1999) Agronomic implications of C_4 photosynthesis, In Sage, R.F. Monson, R.K. (Eds.) *C₄ plant biology*. Academic Press, San Diego, Calif (USA) pp. 473 – 507 ISBN No. 0-12-614440-0

Chojnacka, K. (2010) Biosorption and bioaccumulation – the prospects for practical applications. *Environment International* 36: 299 – 307. ISSN No. 0160 - 4120

Caird, M.A. Richards, J.H. Donovan, L.A. (2007) Nighttime stomatal conductance and transpiration in C_3 and C_4 plants. *Plant Physiology* 143: 4 – 10 ISSN No. 1532 – 2548

Chapin, F.S. (1991) Effects of multiple environmental stresses on nutrient availability and use. In Mooney, H.A. Winner, W. Pell, E.J. (Eds.), *Response of plant to multiple stresses*. Physiological Ecology Series Academic Press, San Diego (USA) 67 – 88 ISBN No. 012505355X

Chen, X. Tu, C. Burton, M.G. Watson, D.M. Burkey, K.O. Hu, S. (2007) Plant nitrogen acquisition and interactions under elevated carbon dioxide: Impact of endophytes and mycorrhizae. *Global Change Biology* 9: 452 – 460 ISSN No. 1365 – 2486

Chen, Y. Shen, Z. Li, X. (2004) The use of vetiver grass (*Vetiveria zizanioides*) in the phytoremediation of soils contaminated with heavy metals. *Applied Geochemistry* 19: 1553 – 1565. ISSN No. 0833 – 2927

Chiu, K.K. Ye, Z.H. Wong, M.H. (2006) Growth of *Vetiveria zizanioides* and *Phragmites australis* on Pb/Zn and Cu mine tailings amended with manure compost and sewage sludge: A green house study. *Bioresource Technology* 97: 158 – 170 ISSN No. 0960 – 8524

Choudhry, M., Jetley, U. K., Khan, M. A., Zutshi, S., Fatma, T. (2006) Effect of heavy metal stress on proline, malondialdehyde, and superoxide dismutase activity in the cyanobacterium *Spirulina platensis*-S5. *Ecotoxicology and Environmental Safety*, 66: 204 – 209.ISSN No. 0147 – 6513

Clemens, S. (2001) Molecular mechanisms of plant metal tolerance and homeostasis. *Planta* 212: 475 – 486.

Doubnerová, V. Ryšlavá, H. (2011) What can enzymes of C_4 photosynthesis do for C_3 plant under stress? *Plant Science* 180: 575 – 583 ISSN No. 0168 – 9452

Dengler, N.G. Nelson, T. (1999) Leaf structure and development in C_4 plants. In: Sage, R.F. Monson, R.K. (Eds.) C_4 plant biology. Academic Press, San Diego, Calif (USA) pp 133 – 172 ISBN No. 0-12-614440-0

Derner, J.D. Johnson, H.B. Kimball, B.A. Pinter Jr., P.J. Polley, H.W. Tischler, C.R. Buttons, T.W. Lomorte, R.L. Wall, G.W. Adam, N.R. Leavitt, S.W. Ottman, M.J. Matthias, A.D. Brooks, T.J. (2003) Above and below-ground responses of C_3 –C_4 species mixtures to elevated CO_2 and water availability. *Global Change Biology* 9: 452 – 460 ISSN No. 1365 – 2486

Dickinson, C.E. Dodd, J.L. (1976) Phenological pattern in the shortgrass prairie *American Midland Naturalist* 96: 367 – 378 ISSN No. 1938 – 4238

Du, S. Fang, S.C. (1982) Uptake of elemental mercury vapor by C_3 and C_4 species. *Environmental and Experimental Botany* 22 (4): 437 – 443 ISSN No. 0098 – 8472

Ebbs, S.D. Kochian, L.V. (1998) Phytoextraction of Zinc by oat (*Avina sativa*), barley (*Hordeum vulgare*) and Indian mustard (*Brassica juncea*). *Environment Science and Technology* 32: 802 – 806 ISSN No. 0013 – 936X

Edwards, E.J. McCaffery, S. Evans, J.R. (2005) Phosphorus status determines biomass response to elevated CO_2 in a legume: C_4 grass community. *Global Change Biology* 11: 1968 – 1981 ISSN No. 1365 – 2486

Ehleringer, J.R. Björkman, O. (1977) Quantum yields for CO_2 uptake in C_3 and C_4 plants: dependence on temperature, CO_2 and O_2 concentration. *Plant Physiology* 59: 86 – 90. ISSN No. 1532 – 2548

Ehleringer, J.R. Cerling, J.E. Helliker, B.R. (1999) C_4 photosynthesis, atmospheric CO_2 and climate. *Oecologia* 112: 285 – 299 ISSN No. 0029 – 8549

Evans, L.T. (1993) Crop evolution, adaptation and yield. Cambridge University Press, Cambridge (UK). ISBN No. 0-521-22571-X

Gupta, A.K. Mishra, R.K. Sinha, S. Lee, B. (2010). Growth, metal accumulation and yield performance of *Brassica campestris* L. (cv. Pusa Jaikisan) grown on soil amended with tannery sludge/fly ash mixture. *Ecological Engineering* 36: 981 – 991. ISSN No. 0925 – 8574

Gaur, A. Adholeya, A. (2004) Prospects of arbuscular mycorrhizal fungi in phytoremediation of heavy metal contaminated soils. *Current Science* 86: 528 – 534. ISSN No. 0011 – 3891

Galloway, J.N. Thornton, J.D. Norton, S.A. Volcho, H.L. McLean, R.A. (1982) Trace metals in atmospheric deposition: a review and assessment. *Atmospheric Environment* 16: 1677 - 1700. ISSN No. 1352 – 2310

Garbisu, C. Alkorta, I. (2001) Phytoextraction: a cost effective plant based technology for the removal of metals from the environment. *Bioresource Technology* 77: 229 – 236. ISSN No. 0960 – 8524

Hikosaka, K. Onoda, Y. Kinugasa, T. Nagashima, H. Anten, N.P.R. Hirose, T. (2005) Plant responses to elevated CO_2 concentration at different scales: leaf whole plant, canopy, and population. *Ecological Research* 20: 243 – 253 ISSN No. 1440 – 1703

Hu, S. Wu, J. Burkey, K.O. Firestone, M.K. (2005) Plant and microbial N acquisition under elevated atmospheric CO_2 in two microcosm experiments with annual grasses. *Global Change Biology* 11: 223 – 233 ISSN No. 1365 – 2486

Hall, J.L. Williams, L.E. (2003) Transitional metal transporters in plants. *Journal of Experimental Botany* 54: 2601 – 2613 ISSN No. 0022 – 0957

Hinchman, R.R. Nergi, M.C. Gatliff, E.G. (1998) Phytoremediation: using green plants to cleanup contaminated soil, ground water and wastewater.
http://www.treemediation.com/technical/phytoremediation_1998.pdf.

Jonak, C. Nakagami, H. Hirt, H. (2004) Heavy metal stress. Activation of distinct mitogen – activated protein kinase pathways by Copper and Cadmium. *Plant Physiology* 136: 3276 – 3283 ISSN No. 1532 – 2548

Jones, P. Kortenkamp, A. O'Brien, P. Wang, G. Yang, G. (1991) Evidence for the generation of hydroxyl radicals from a chromium (V) intermediate isolated from the reaction of chromate with glutathione. *Biochimica et Biophysica Acta* 286: 652 – 655 ISSN No. 0006 – 3002

Khade, S.W. Adholeya, A. (2008) Arbuscular mycorrhizal association in plants growing on metal-contaminated and noncontaminated soils adjoining Kanpur tanneries, Uttar Pradesh, India. *Water Air Soil Pollution* 202 (1-4): 45 – 56. ISSN No. 0049 – 6979

Kamal, M. Ghaly, A.E. Mahmoud, N. Côté, R. (2004) Phytoaccumulation of heavy metals by aquatic plants. *Environment International* 29: 1029 – 1039. ISSN No. 0160 – 4120

Khan, A.G. Chaudhry, T.M. Hayes, W.J. Khoo, C.S. Hill, L. Fernandey, R. Gallardo, P. (1998) Physical, chemical and biological characterization of a steelworks waste site at port Kembla, NSW, Australia. *Water, Air Soil Pollution* 104: 389 – 402. ISSN No. 0049 – 6979

Kimball, B. A. (1983) Carbon dioxide and agricultural yield: an assemblage and analysis of 430 prior observations. *Agronomy Journal* 75, pp 779 – 788. ISSN No. 0002 – 1962

Krupa, S. V. (1996) The role of atmospheric chemistry in the assessment of crop growth and productivity. In: Yunus, M. and Iqbal, M. (Eds.) *Plant Response to Air Pollution*, 35. John Wiley & Sons, Chichester, U.K. ISBN No. 0-471-96061-6

Kumar, N.P.B.A. Dushenkov, V. Motto, H. Raskin, I. (1995) Phytoextraction: the use of plants to remove heavy metals from soils. *Environment Science and Technology* 29: 1232 – 1238. ISSN No. 0013 – 936X

Kunze, R. Frommer, W.B. Flügge, U.I. (2001) Metabolic engineering in plants: the role of membrane transport. *Metabolic Engineering* 4: 57 – 66. ISSN No. 1096 – 7176

Lasat, M.M. (2002) Phytoextraction of toxic metals: a review of biological mechanisms. *Journal of Environmental Quality* 31(1): 109 – 120. ISSN No. 0047 – 2425

Leegood, R.C. Edwards, G.E. (1996) *Photosynthesis and the environment*, Vol. 5 Kluwer Academic Publishers, Dordrecht, The Netherlands ISBN No. 0792361431

Lloyd, J. Farquhar, G.D. (1994) ^{13}C discrimination during CO$_2$ asimilation by the terrestrial biosphere. *Oecologia* 99: 201 – 215 ISSN No. 0029 – 8549

Ma, J.F. Furukawa, J. (2003). Recent progress in the research of external Al detoxification in higher plants: a mini review, *Journal of Inorganic Biochemistry* 97: 46-51.ISSN No. 0162 – 0134

Matsuoka, M. Hata, S. (1987) Comparative studies of phosphoenolpyruvate carboxylase from C$_3$ and C$_4$ plants. *Plant Physiology* 85: 947 – 951 ISSN No. 1532 – 2548

Matsuoka, M. Nomura, M. Agarie, S. Miyao-Tokutomi, M. Ku, M.S.B. (1998) Evolution of C$_4$ photosynthetic genes and over expression of maize C$_4$ gene in rice. *Journal of Plant Research* 111: 333 – 337 ISSN No. 1618 – 0860

Monz, C.A. Hunt, H.W. Reeves, F.B. Elliott, E.T. (1994) The response of mycorrhizal colonization to elevated CO_2 and climate change in *Pascopyrum smithii* and *Bouteloua gracilis*. *Plant and Soil* 165: 75 – 80 ISSN No. 0032 – 079X

Nieboer, E. Richardson, D.H.S. (1980) The replacement of the non descript term 'heavy metal' by a biologically significant and chemically significant classification of metal ions. *Environmental Pollution* B1: 3 – 26 ISSN No. 0269 – 7491

Nieboer, E. Richardson, D.H.S. (1990) The replacement of the nondescript term "heavy metals" by a biologically and chemically significant classification of metal ions. *Environmental Pollution* B 1: 3 – 26. ISSN No. 0269 – 7491

Niu, S. Zhang, Y. Yuan, Z. Liu, W. Huang, J. Wan, S. (2006) Effects of interspecific competition and nitrogen seasonally on the photosynthetic characteristics of C_3 and C_4 grasses. *Environmental and Experimental Botany* 57: 270 – 277 ISSN No. 0098 – 8472

O'Connor, P.J. Smith, S.E. Smith, E.A. (2002) Arbuscular mycorrhizas influence plant diversity and community structure in a semiarid herbland. *New Phytologist* 154: 209 – 218 ISSN No. 1469 – 8137

O'Leary. M.H. (1982) Phosphoenolpyruvate carboxylase: an enzymologist's view. *Annual Reviews in Plant Physiology* 33: 279 – 315 ISSN No. 0066 – 4294

Oren, R. Ellsworth, D.S. Johnsen, K.H. Phillips, N. Ewers, B.E. Maier, C. Schafer, K.V.R. McCarthy, H. Hendrey, G. McNulty, S.G. Katul, G.G. (2001) Soil fertility limits carbon sequestration by forest ecosystems in a CO_2 enriched atmosphere. *Nature* 411: 469 – 472 ISSN No. 0028 – 0836

Pulford, I.D. Watson, C. (2003) Phytoremediation of heavy metal – contaminated land by trees – a review. *Environment International* 29: 529 – 540. ISSN No. 0160 – 4120

Patterson, D.T. (1995) Effects of environmental stress on weed/crop interactions. *Weed Science* 43: 483 – 490. ISSN No. 0043 – 1745

Pickering, W.F. (1995) General strategies for speciation. In Ure, A.M. Davidson, C.M. (Eds.) *Chemical Speciation in the Environment* Blackie Academic and Professional Press. London (UK) 9 – 31 ISBN No. 0-632-05848-X

Poorter, H. (1993) Interspecific variation in growth response of plants to an elevated ambient CO_2 concentration. *Vegetation* 104/105: 77 – 97

Rajendrudu, G. Das, V.S.R. (1982) Biomass production of two species of cleome exhibiting C_3 and C_4 photosynthesis. *Biomass* 2: 223 – 227 ISSN No. 0960 – 8524

Riesterer, J.L. Casler, M.D. Undersander, D.J. Combs, D.K. (2000) Seasonal yield distribution of cool season grasses following winter defoliation. *Agronomy Journal* 92: 974 – 980 ISSN No. 0002 – 1962

Rillig, M.C. Allen, M.F. (1998) Arbuscular mycorrhizae of Gutierrezia sarothrae abd elevated carbon dioxide: evidence for shift in C allocation to and within the mycobiont. *Soil Biology and Biochemistry*, 30: 2001 – 2008. ISSN No. 0038 - 0717

Rillig, M.C. Allen, M.F. Klironomos, J.N. Chiariello, N.R. Field, C.B. (1998) Plant species – specific changes in root inhabiting fungi in a California annual grassland: responses to elevated CO_2 and nutrient. *Oecologia* 113: 252 – 259. ISSN No. 1432 – 1939

Srivastava, J. Kayastha, S. Jamil, S. Srivastava, V. (2008) Environmental perspective of *Vetiveria zizanioides* (L.) Nash. *Acta Physiologia Plantarum* 30: 413 – 417 ISSN No. 1861 – 1664

Response of C₃ and C₄ Plant Systems Exposed to Heavy Metals for Phytoextraction at Elevated Atmospheric
CO₂ and at Elevated Temperature

35

Sage, R.F. (2004) The evolution of C₄ photosynthesis. *New Phytologist* 161: 341 – 370. ISSN No. 1469 – 8137

Schützendübel, A. Polle, A. (2002) Plant responses to abiotic stresses: heavy metal-induced oxidative stress and protection by mycorrhization. *Journal of Experimental Botany* 53 (372): 1351 – 1365. ISSN No. 0022 – 0957

Sage, R.F. Pearcy, R.W. (2000) The physiological ecology of C₄ photosynthesis. In: Leegood, R.C. Sharkey, T.D. vonCaemmerer, S. (Eds.) *Photosynthesis: Physiology and Metabolism* Kluwer Academic Publishers, Dordrecht (The Netherlands) 497 – 532 ISBN No. 0792361431

Sage, R.F. Wedin, D.A. Li, M.R. (1999) The biogeography of C₄ photosynthesis: patterns and controlling factors. In: Sage, R.F. Monson, R.K. (Eds.) *C₄ plant biology.* Academic Press, San Diego, Calif (USA) 313 – 373 ISBN No. 0-12-614440-0

Srivastava, A.K. Purnima, X. (1998) Phytoremediation for heavy metals – a land plant based sustainable strategy for environmental decontamination. *Proceddings of National Academy of Science, India, Section B Biological Science* 68 (3-5): 199 – 215. ISSN No. 0073 – 6600

Staddon, P.L. Fitter, A.H. (1998) Does elevated atmospheric carbon dioxide affect arbuscular mycorrhizas? *Trends in Ecological Evolution* 13: 455 – 458. ISSN No. 0169 – 5347

Shi, X. Dalal, N.S. Kasprzak, K.S. (1993) Generation of free radicals from hydrogen peroxide and lipid hydroperoxides in the presence of Cr(III). *Biochimica et Biophysica Acta* 302: 294 – 299. ISSN No. 0006 – 3002

Stohs, S.J. Bagchi, D. (1995) Oxidative mechanisms in the toxicity of metal ions. *Free Radical Biology and Medicine* 18: 321 – 336 ISSN No. 0891 – 5849

Suresh, B. Ravishanker, G.A. (2004) Phytoremediation – a novel and promising approach for environmental cleanup. *Critical Reviews in Biotechnology* 24: 97 – 124 ISSN No. 0738 – 8551

Tang, J. Chen, J. Chen, X. (2006) Responses of 12 weedy species to elevated CO₂ in low phosphorus availability soil. *Ecological Research* 21: 664 – 670 ISSN No. 0912 – 3814

Tang, J. Xu, L. Chen, X. Hu, S. (2009) Interaction between C₄ barnyard grass and C₃ upland rice under elevated CO₂: Impact of mycorrhizae. *Acta Oecologica* 35: 227 – 235 ISSN No. 1146 – 609X

Tolbert, N.E. (1971) Microbodies-peroxysomes and glyoxysomes. *Annual Reviews in Plant Physiology* 22: 45 – 74 ISSN No. 0066 – 4294

Treseder, K.K. (2004) A meta analysis of mycorrhizal responses to nitrogen, phosphorus, and atmospheric CO₂ in field studies. *New Phytologist* 164: 347 – 355 ISSN No. 1469 – 8137

Truong, P.N. Baker, D. (1998) Vetiver grass system for environmental protection. *Technical Bulletin 1998/1 – Pacific Rim Vetiver Network,* Bangkok, Thailand

Tu, S. Ma, L.Q. (2003) Interactive effects of pH, arsenic and phosphorous on uptake of As and P and growth of the arsenic hyperacuumulator *Pteris vittata* L. under hydroponic conditions. *Environmental and Experimental Botany* 50: 243 – 251. ISSN No. 0098 – 8472

Ueno, O. (2001) Environmental regulation of C₃ and C₄ differentiation in the amphibious sedge *Eleocharis vivipara. Plant Physiology* 127: 1524 – 1532 ISSN No. 0032 – 0889

Wand, S.J. Midgley, G.F. Jones, M.H. Curtis, P.S. (1999) Response of wild C₄ and C₃ (Poaceae) species to elevated atmospheric CO₂ concentartion: a meta analytic test of current theories and perceptions. *Global Change Biology* 5: 723 – 741 ISSN No. 1365 – 2486

Weast, R.C. (1984) *CRC Handbook of Chemistry and Physics, 64th edn.* CRC Press, Boca Raton ISBN No. 0849307406

Weber, A.P.M. Caemmerer, S. V. (2010) Plastid ransport and metabolism of C_3 and C_4 plants comparative analysis and possible biotechnological exploratation. *Current Opinion in Plant biology* 13: 257 – 265 ISSN No. 1369 – 5266

Wilde, E.W. Brigmon, R.L. Dunn, D.L. Heitkamp, M.A. Dagnan, D.C. (2005) Phytoextraction of lead from firing range soil by vetiver grass. *Chemosphere* 61: 1451 – 1457. ISSN No. 0045 – 6535

Wilson, G.W.T. Hartnett, D.C. (1998) Interspecific variations in plant response to mycorrhizal colonization in tall grass prairie. *American Journal of Botany* 85: 1732 – 1738 ISSN No. 0002 – 9122

Winslow, J.C. Hunt Jr., E.R. Piper, S.C. (2003) The influence of seasonal water availability on global C_3 versus C_4 grassland biomass and its implications for climate change research. *Ecological Modelling* 163: 153 – 173 ISSN No. 0304 – 3800

Ye, Z.H. Baker, A.J.M. Wong, M.H. Willis, A.J. (2003) Copper tolerance, uptake and accumulation by *Phragmites australis*. *Chemosphere* 50: 795 – 800 ISSN No. 0045 – 6535

Zheng, W. J., Zheng, X. P. and Zhang, C. L. (2000). A survey of photosynthetic carbon metabolism in 4 ecotypes of *Phragmites australis* in northwest China: Leaf anatomy, ultra-structure, and activities of ribulose 1,5-biphosphate carboxylase, phosphoenolpyruvate carboxylase and glycollate oxidase. *Physiologia Plantarum* 110: 201–208. ISSN No. 1399 - 3054

3

Plants and Soil Contamination with Heavy Metals in Agricultural Areas of Guadalupe, Zacatecas, Mexico

Osiel González Dávila,[1] Juan Miguel Gómez-Bernal[2]
and Esther Aurora Ruíz-Huerta[2]
[1]*The University of London – School of Oriental and African Studies (SOAS)*
[2]*Posgrado en Ciencias de la Tierra, Instituto de Geofísica*
Universidad Nacional Autónoma de México (UNAM)
[1]*United Kingdom*
[2]*México*

1. Introduction

The environmental impact of mine tailings has been largely documented around the world. Deterioration and contamination of soils, groundwater and superficial water as well as alterations in the hydrological systems have been associated with mining wastes (Figueroa *et al* 2010). Heavy metal contamination of plants, soil and water affects several countries worldwide posing a serious threat to the health of millions of people. Due to its long mining history, Mexico is among the most affected countries by this serious environmental problem.

A geochemical comparative study was conducted in the municipality of Guadalupe in Zacatecas, Mexico. The objectives were to measure the bioconcentration factor in plants of agronomic interest in function of their heavy metal absorption, to identify the toxicity order of heavy metals in plants of agronomic interest, to assess potential environmental impacts taking into account the particularities of the selected crop and to evaluate the potential consequences on the region's food security.

Zacatecas state is located in north central Mexico (see figure 1). There, metallic ores are abundant and diverse. The state has 450 years of mining tradition with the consequent accumulation of mining tailings (Salas–Luévano *et al* 2009). Currently, Zacatecas state is the most important silver producer in Mexico. During the year 2010, 1,855,145 kilograms of silver were produced in Zacatecas (INEGI 2011). Amalgamation for silver extraction, also known as patio process, consists in adding mercury to the silver ore in order to obtain a silver amalgam as the final product. Amalgamation was used extensively throughout the period from 1570 to 1820. Most of the heavy metals lost via amalgamation were carried by rivers and deposited in the plain areas of the Zacatecan valley in what is now the Guadalupe municipality. Most of these areas are currently used for crop farming since there are no restrictions imposed by the Mexican authorities (Santos–Santos *et al* 2006).

Previous studies have found high levels of Pb, As, Hg and F⁻ in groundwater extraction wells that supply Guadalupe municipality (Leal & Gelover 2002; Castro *et al* 2004; González Dávila 2011). In addition to drinking water health risks, there is also risk of potential levels of heavy metals entering the food chain via absorption by crops from contaminated soil and water. Heavy metal contaminated crops could aggravate human health risk when consumed along with heavy metal contaminated drinking water (Brammer 2008; Duxbury 2007; Santos–Santos *et al* 2006).

The most important staple food in Mexico is maize (*Zea mays* L.). Meals are based on maize, with tortillas providing much of the caloric intake both in rural and urban areas. Due to its importance for the food security of the region, it was decided to analyse the contents of heavy metals in maize plants.

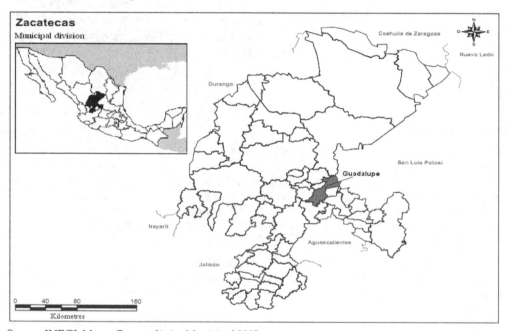

Source: INEGI. Marco Geoestadístico Municipal 2005

Fig. 1. Location of Guadalupe municipality in Zacatecas State

2. Methods and materials

2.1 Geographical delimitation

Soil and maize plant samples were collected from agricultural areas from the municipality of Guadalupe, Zacatecas. Guadalupe municipality is located at an average 2,280 metres above sea level (lat 22° 45' N and long 102° 31' O). The site is characterized by a climate arid sub tropic tempered, throughout the year. The average annual temperaturee is 17°C. The average maximum temperature is 30°C and occurs during May. The average minimum temperature is 3°C and occurs in January. There is an average annual precipitation of 463 mm (CONAGUA 2010:25). Due to Zacatecas' climate and environmental conditions, irrigation is very important in maize production. According to the Service of Agrifood and Fishery Information (SIAP 2010), in Zacatecas 34,918 hectares of land producing maize were irrigated during 2010. Soil and plant samples were collected from 5 different irrigation zones in the municipality during June 2011. In the southern part of the municipality, samples were collected from agricultural land in Noria Blanca and Las Mangas. In the central part, samples were collected from La Zacatecana and in the northern part samples were collected from agricultural land in Osiris. The coordinates of sampling points can be found in table 1. The map in figure 2 shows each of the collection points.

	Zone	Alt	lat N	long W
1	La Noria	2267	22° 40′ 02.1″	102° 28′ 52.5″
2	Las Mangas	2254	22° 41′ 46.5″	102° 29′ 28.1″
3	La Zacatecana a	2220	22° 44′ 37.2″	102° 27′ 55.5″
4	La Zacatecana b	2223	22° 44′ 32.1″	102° 28′ 09.6″
5	Osiris	2190	22° 46′ 02.2″	102° 26′ 56.8″

Table 1. Sampling points coordinates

Source: Mapped with Google Earth using the authors' data

Fig. 2. Plants and soil sampling points

2.2 Sample size

The number of samples n was calculated with the formula: $n = [Za^2 * p * (1-p)] / d^2$. A 95% confidence level was established and a Za of 1.96 was obtained. Following Santos-Santos *et al* (2006), a proportion value p of 0.05 and a precision factor d of 8.5% were selected. Thus, the number of samples n was calculated as 25.26 samples. Thus, it was decided to collect five maize plant and soil samples in a 100 m² area of agricultural land in each irrigation zone. Nevertheless, 2 extra samples were collected because mine tailings were found next to agricultural land in zone 3.

2.3 Soil samples analysis

Soil pH was measured in soil-H_2O suspension (1:2.5, w/w) and electrical conductivity was measured in a 1:5 soil to water suspension using an HI 9828 Multiparameter portable (HANNA instruments) with intelligent probe and T.I.S. Total N was determined using the Kjeldhal method (Black 1965). Organic matter content was determined by the Walkley and Black procedure (Nelson & Sommers 1982). Available P was measured colorimetrically by the molybdenum blue method (Olsen & Sommers 1982). Soil samples were dried at 60 °C for 75 h; then each sample was crushed, sieved (< 325 μm), homogenized, and weighed. Soil particle size distribution was measured using the hydrometer method (Allen et al 1974). Carbonate content was determined following Horton & Newson (1953) methodology. Available (DTPA-extractable) heavy metal concentrations (Pb, Cd, Fe, Cu and Mn) were determined by atomic absorption spectrophotometry. Total heavy metals in soil and plant samples were measured by energy-dispersive X-ray fluorescence spectrometry, using a NITON XL3t of Thermo Fisher Scientific. X-ray spectra were analyzed with Niton Data Transfer software suite. The spectrometer was calibrated for heavy metals using certified standards from NIST (National Institute of Standards and Technology) Montana soil 2711 and 2710a and peach in plants. Intermediate and high heavy metal concentration standards, traceable to NIST, were prepared in our facility to have a wide range calibration curve. Heavy metals concentrations in the samples were measured three times. Arsenic (a metalloid) in soils and plants samples was also determined using energy-dispersive X-ray fluorescence. This technique has been accepted by the U.S. Environmental Protection Agency to measure arsenic in dry solid samples (Melamed 2004:4).

2.4 Plant samples analysis

Plant samples were collected from the top layer (0-30 cm) of agricultural land. They contained a mix of spoil and soil. Thus, samples were washed thoroughly in the laboratory with running tap water, followed by three rinses with deionized water (18 MΩcm-1, Milli-Q Millipore) and a rinse of tri-distilled water. All plant samples were carefully divided into shoots and roots. They were dried at 60°C for 75h. The oven-dried plant samples were then crushed, sieved (< 325 μm), homogenized, and weighed. Later, arsenic and heavy metal concentrations were determined by energy dispersive X-ray fluorescence. The translocation factor (TF) for metals within a given plant was calculated as metal concentration in shoot divided by that in root (Tu et al 2003; Rizzi et al 2004). The bioconcentration factor (BFC) was expressed by the ratio of metal concentration in plant above ground part to total metal concentration in soil (Rotkittikhun et al 2006).

3. Results

3.1 Soil samples results

Table 2 shows the results of total concentrations for the following elements: Pb, As, Hg, Zn, Cu, Fe, Mn and K. All the results are expressed in ppm. Tests for Cd, Ag and Ni were also conducted but the concentration levels were in all samples under the limit of detection. In zones 1 to 4, at least five soil samples were collected. In zone 5, three soil samples were collected. Sample 17 corresponds to a tailing sample collected in zone 3 from a tailing pond located next to agricultural land (see figure 3).

Zone	Sample	Pb	As	Hg	Zn	Cu	Fe	Mn	K
1 La Noria	zac-1	28.69	< BDL	< BDL	72.78	< BDL	21828.6	481.24	11771.6
	zac-2	26.02	< BDL	< BDL	74.26	< BDL	23137.2	506.34	11193
	zac-3	29.13	< BDL	< BDL	76.33	< BDL	23575.2	581.25	12148.7
	zac-4	26.17	< BDL	< BDL	87.78	< BDL	21849.3	494.12	8126.95
	zac-5	23.34	< BDL	< BDL	69.91	< BDL	23030	519.22	10738.2
	zac-6	22.60	< BDL	< BDL	73.64	< BDL	21620.9	495.55	8555.12
2 Las Mangas	zac-7	52.04	15.92	< BDL	109.4	< BDL	19778.2	527.64	14058.3
	zac-8	28.44	12.91	< BDL	64.54	< BDL	18523.3	622.17	12772.8
	zac-9	20.93	< BDL	< BDL	72.4	< BDL	19323.8	365.46	12529.6
	zac-10	27.25	13.14	< BDL	70.06	< BDL	19240.8	560.97	14183.9
	zac-11	36.43	14.06	< BDL	98.65	< BDL	22773.7	525.42	11743.6
3 La Zacatecana A	zac-12	534.94	87.4	16.69	997.27	95.41	36453.5	928.08	13191.9
	zac-13	660.34	163.34	20.73	1392.47	113.17	37976.4	927.29	11631.4
	zac-14	644.52	143.82	18.58	1233.23	114.26	38234.1	895.54	12067.5
	zac-15	518.84	85.53	< BDL	882.36	105.6	35873	812.61	13470.3
	zac-16	552.36	94.51	< BDL	946.45	95.36	35726.3	914.2	11438.5
	zac-17*	5660.25	289.9	505.9	10086.5	1323.82	55330.6	1792.39	10466.4
4 La Zacatecana B	zac-18	540.39	70.64	< BDL	889.6	113.49	34742.6	783.97	12821.7
	zac-19	572.71	68.82	21.92	955.41	107.81	35413.9	859.98	11564
	zac-20	661.17	90.95	17.9	1110.74	146.63	36420.7	945.89	14131.8
	zac-21	634.74	59.41	25.27	1049.17	136.21	36023.1	818.63	12606.4
	zac-22	625.63	77.22	25.51	1042.01	145.5	39214	824.22	12492.4
	zac-23	639.82	68.28	20.5	982.33	132.9	35100	749.13	11480.9
	zac-24	540.39	70.64	37.69	1303.11	147.76	35619.1	1189.53	12795.9
5 Osiris	zac-25	105.58	< BDL	< BDL	182.35	56.2	40001.9	780.22	9100.29
	zac-26	88.65	21.39	< BDL	161.41	55.2	40888.1	707.06	9434.11
	zac-27	105.96	< BDL	< BDL	188.72	64.5	44801.5	816.88	9836.79

Results for Pb, As, Hg, Zn, Cu, Fe, Mn and K are in ppm. Areas were defined according to fig. 2
* Sample 17 corresponds to a mine tailing sample collected in zone 3.
< BDL = Below detection limit.

Table 2. Arsenic and heavy metal concentrations (ppm) in agricultural soils of Guadalupe, Zacatecas.

Table 3 shows the pH and electrical conductivity in the soil samples. The samples collected in zones 1, 2 and 5 are moderately alkaline. Soil samples from zones 3 and 4 are slightly alkaline. The organic matter was also determined. It should be noted that the percentage of organic matter in soils samples collected in zones 3 and 4 is higher than in the rest. This can be explained by the fact that wastewater irrigation is a common practice in that specific area. The other parameters shown in the table are total nitrogen, phosphorous and calcium carbonate. The high levels of phosphorus and total nitrogen found in samples from zones 3 and 4 are congruent with the levels of organic matter found. Table 4 shows the available (DTPA-extractable) heavy metal concentrations in ppm. It should be noted that the availability of such elements is higher in zones 3 and 4. This is congruent with the information shown in table 2.

Zone	pH	EC dS/m	OM %	TN %	P (ppm)	CaCO$_3$T %
1 La Noria	8.288	1.206	1.294	0.062	20.55	1.814
2 Las Mangas	7.872	0.538	1.628	0.127	16.08	0.602
3 La Zacatecana A	7.686	3.570	3.868	0.192	59.10	3.058
4 La Zacatecana B	7.602	2.704	4.062	0.208	74.71	3.152
5 Osiris	8.304	1.678	1.494	0.072	8.48	1.458

Areas were defined according to fig. 2. EC=Electrical Conductivity, OM= Organic Matter, TN = Total Nitrogen.

Table 3. Chemical analysis of agricultural soils of Guadalupe, Zacatecas.

Zone	Pb	Cd	Fe	Cu	Mn
1 La Noria	0.562	0.013	3.560	0.588	17.380
2 Las Mangas	2.314	0.086	8.664	1.344	29.844
3 La Zacatecana A	67.940	5.036	74.228	35.808	23.440
4 La Zacatecana B	85.656	4.284	33.306	29.688	26.810
5 Osiris	4.138	0.574	3.528	5.546	15.214

Results for Pb, Cd, Fe, Cu and Mn are in ppm. Areas were defined according to fig. 2

Table 4. Available (DTPA-extractable) heavy metals in agricultural soils of Guadalupe, Zacatecas.

Fig. 3. Tailing pond close to agricultural land in Guadalupe, Zacatecas.

3.2 Plant samples results

The concentrations of Pb, As, Zn, Cu, Fe and Mn in the roots and shoots of maize plants collected in the study area are summarized in table 5. In zones 1 to 4, at least five maize plant samples were collected. In zone 5, three maize plant samples were collected. Tests for Cd, Hg, Ag and Ni were also conducted. However, the levels of those elements were under the limit of detection in all samples.

Zone	Samples	Pb	As	Zn	Cu	Fe	Mn
1	Roots	NA	15.26	153.97	74.45	11174.63	295.83
	Shoots	NA	NA	94.38	69.87	766.66	324.03
2	Roots	NA	NA	31.13	42.75	1053.51	NA
	Shoots	NA	NA	89.02	81.58	956.74	271.69
3	Roots	293.24	98.15	849.74	111.49	25359.64	629.71
	Shoots	21.39	NA	688.63	121.71	2196.75	150.1
4	Roots	79.77	44.14	462.5	213.63	11357.48	223.24
	Shoots	16.8	NA	438.07	120.35	1565.75	263.69
5	Roots	18.55	NA	236.69	89.84	13233.08	318.83
	Shoots	NA	NA	177.69	104.57	1992.08	485.14

Table 5. Arsenic and heavy metal concentrations (ppm) in roots and shoots of maize plants collected in Guadalupe, Zacatecas. NA=Not available

3.3 Bioconcentration and translocation factors in plant samples

Table 6 shows the bioconcentration and translocation factors for metals in maize plant samples. The toxicity order is discussed in section 4.2.

Zone	Factor	Pb	As	Zn	Cu	Fe	Mn
1	BCF	NA	NA	2.02	NA	0.49	0.57
	TF	NA	NA	0.61	0.94	0.07	1.1
2	BCF	NA	NA	0.38	NA	0.05	NA
	TF	NA	NA	2.86	1.91	0.91	NA
3	BCF	0.5	0.85	0.78	1.06	0.69	0.7
	TF	0.07	NA	0.81	1.09	0.09	0.24
4	BCF	0.13	0.6	0.44	1.64	0.32	0.24
	TF	0.21	NA	0.95	0.56	0.14	1.18
5	BCF	0.19	NA	1.33	1.53	0.32	0.42
	TF	NA	NA	0.75	1.16	0.15	1.52

Table 6. Bioconcentration and translocation factors. NA=Not available

4. Discussion

4.1 Soil contamination

The Mexican Official Norm NOM-147-SEMARNAT/SSA1-2004 (SEMARNAT 2007) established the following guideline values for arsenic and heavy metals in agricultural soil in 2007:

Element	Guideline value (ppm)
As	22
Cd	37
Hg	23
Ag	390
Ni	1600
Pb	400

Table 7. Mexican guideline values for arsenic and heavy metals in agricultural soil

This study has identified arsenic, lead and mercury contamination in agricultural soil from Guadalupe, Zacatecas (see table 2). Table 8 presents the mean, standard deviation (SD) and range of Pb, As and Hg concentrations found in the five sampling zones. Zones 3 and 4 located in La Zacatecana are the most contaminated. All the soil samples collected in those areas are above the 400 ppm maximum allowed level of Pb in soils established by the Mexican Official Norm. Although Pb concentrations are lower than those reported in other mining regions in Mexico (see for example Gutiérrez-Ruiz et al 2007 that report a Pb range of 972-16,881 ppm), the Pb contamination levels are unquestionably high and toxic. Arsenic concentrations were also high in the studied areas -ranging from 15.92 to 163.34 ppm- even compared to those reported in other mining regions from Mexico (see for example Mendoza-Amézquita et al. 2006 that report As concentrations of 21-36 ppm) and North America (Moldovan et al. 2003 report As concentrations of 56-6,000 ppm). It should be noted that all the samples in zones 3 and 4 are above the 22 ppm As guideline. However, As concentrations found in this study were low compared to concentrations reported by Méndez & Armienta (2003) in Zimapan, Hidalgo, Mexico (2,550-14,600 ppm) and Ortega-Larrocea et al (2009) for the same area (up to 2,869 ppm). On the other hand, three soil samples from zone 4 were above the 23 ppm Hg guideline. Hg contamination is not evident in zones 1,2 and 5. Table 9 shows that there is a strong positive correlation between the presence of Pb and As in soils of the region. The correlation is significant at the 0.01 level. This relationship suggests a common source of contamination and it is very likely that it is related to the same kind of mining activities. It is very important to mention that during the fieldwork the authors found that a local mining company dug a tailing pond and was filling it with mining waste just 13 meters away from agricultural land in zone 3 "La Zacatecana A" (see figure 3). High heavy metal levels were found in the tailing sample collected there.

In their exploratory study, Santos–Santos et al (2006) reported that the main source of heavy metal contamination in Guadalupe's soil is related to old mining activities carried out in the surrounding area of Osiris and La Zacatecana. However, it was found that new mine tailings in the area are recklessly managed and there is an alarming lack of enforcement mechanisms to oblige the mining companies to obey the environmental laws and regulations. Those new tailings are undoubtedly a source of heavy metal contamination of the neighbouring agricultural land. Although Manzanares et al (2003) reported normal levels of lead and mercury in blood of sampled people at La Zacatecana, it is very likely that those concentrations have increased over time. Two heavy metal exposition routes can be identified. In the first place, there is a respiratory intake of particles and dust from contaminated soil. Second, as it is explained in the following section, there is a deposition of heavy metals in crops aimed for human consumption. Therefore, a blood study should be conducted again among the people of the region. Due to the presence of the new mine tailings in the region, a higher exposure to heavy metals is expected.

Zone	n	Pb				As				Hg*			
		Mean	SD	Range	% > 400 ppm Pb	Mean	SD	Range	% > 22 ppm As	Mean	SD	Range	% > 23 ppm As
1	6	25.99	2.67	22.60 - 29.13	0	<BDL	NA	< BDL	0	< BDL	NA	< BDL	0
2	5	33.02	11.98	20.93 - 52.04	0	14.01	1.37	12.91 - 15.92	0	< BDL	NA	< BDL	0
3	5	582.20	65.44	518.84 - 660.34	100	114.92	36.12	85.53 - 163.34	100	18.67	2.02	16.69- 18.58	0
4	7	602.12	50.02	540.39 - 661.17	100	72.28	9.76	59.41 - 90.95	100	24.80	6.95	17.9-37.69	43
5	3	100.06	9.89	88.65 - 105.96	0	21.39	NA	21.39	0	< BDL	NA	< BDL	0

Results for Pb, As and Hg are in ppm. Areas were defined according to Figure 2. < BDL = Below detection limit. NA=Not available. * For the mean and SD calculation for Hg samples under the limit of detection were excluded.

Table 8. Mean Lead, Arsenic and Mercury levels in soil samples collected in five different risk areas of Guadalupe, Zacatecas

		Pb	As
Pb	Pearson Correlation	1	.865**
	Sig. (2-tailed)		.000
	N	27	18
As	Pearson Correlation	.865**	1
	Sig. (2-tailed)	.000	
	N	18	18

**. Correlation is significant at the 0.01 level

Table 9. Correlation levels of Lead and Arsenic

4.2 Plants contamination

Soils from zones 3 and 4 have the highest levels of heavy metal concentrations (see tables 2 and 4). High levels of heavy metals were also found in plants collected in those areas. Plants from zones 1 and 2 showed lower heavy metal concentrations than plants from the other 3 zones. Toxic levels of Cu and Zn in plants were found in all zones except zone 2. One of the objectives of this study was to measure the bioconcentration and the transference factor in maize plants and to indicate the toxicity order of heavy metals in the plants. It was found that the amount of metals was higher in roots and shoots of plants growing in the most contaminated soil of zones 3 and 4. Consistently with other works (Bidar et al 2007; Marques et al 2009), heavy metal accumulation occurred more frequently in roots than in shots. The BCF shows that there is a higher accumulation of Zn and Cu in maize plants. In some zones, Zn concentration exceeded by two of times the critical limits proposed by Kabata-Pendias (2001) (see table 10). The BCF factor in zone 1 was 2.02 for Zn, followed by 0.57 for Mn and 0.49 for Fe. The other metals were not detected. In zone 2 the BCF was 0.38 for Zn and 0.05 for Fe. In zone 3, the BCF was Cu>As>Zn>Mn>Fe>Pb. The BCF in zone 4 was Cu>As>Zn>Fe>Mn>Pb. And in zone 5, the BCF was Cu>Zn>Mn>Fe>Pb>As. The order of the sampled sites in relation to their BCF from lower to higher is: zone 2 < zone 1 < zone 5 < zone 4 < zone 3.

In regards to the translocation factor (TF), in zone 1 it was Mn>Cu>Zn>Fe>Pb>As. In zone 2 it was Zn>Cu>Fe>Mn>Pb>As. In zone 3 it was Cu>Zn>Mn>Fe>Pb>As. The TF in zone 4 was Mn>Zn>Cu>Pb>Fe>As and in zone 5 it was Mn>Cu>Zn>Fe>Pb>As. The TF shows a higher concentration of Cu, Zn and Mn in the shoots of maize plants and a lower concentration of Pb and As.

Element	Ranges of toxic concentrations in plants (ppm)
Pb	30-300
As	5-20
Hg	1-3
Zn	100-400
Cu	20-100
Mn	400-1000

Source: Kabata-Pendias (2001)

Table 10. Ranges of heavy metals reported to be toxic for plants

4.3 Implications for food security

According to the FAO (2003), **food security** exists when all people, at all times, have physical, social and economic access to sufficient, safe and nutritious food which meets their dietary needs and food preferences for an active and healthy life. The very fact that new mine tailings have been found next to agricultural land in the study area should be considered as a threat to food safety. It is very important to highlight that in zones 3 and 4 the accumulation of Pb and As in plants is very high. Those metals are highly toxic and could be bioaccumulated and transferred to the food chain. This is of particular relevance because of the potential adverse effects on health and food security of people in the region. A second reason of concern is the high levels of Mn, Zn and Cu found both in soils and plants. It should be noted that Cu and Zn are not considered toxic for humans but are toxic for plants and for this reason some countries have posed restrictions to their concentrations in soil. This is of particular relevance because a higher concentration of these elements hinders the development of plants and could reduce land productivity and access to food.

5. Conclusions

The aim of this research was to provide an assessment of heavy metal contamination in five agricultural zones of the Guadalupe municipality in Zacatecas, Mexico. High levels of arsenic, lead and mercury contamination in agricultural soil were found in two irrigation zones. High levels of Zn and Cu were found both in soils and plants in all the areas. Heavy metal absorption in maize plants aimed for human consumption was calculated using the bioconcentration and the translocation factors. The accumulation of Pb and As in plants was very high. Those metals are highly toxic and could be bioaccumulated and transferred to the food chain. Further, high levels of Zn and Cu were found both in soils and plants. Although they are not considered toxic for humans, they are toxic for plants. Several studies have found that high concentrations of these elements hinder the development of plants and could reduce land productivity. A strong and positive correlation of concentration of arsenic and lead in soil suggests that there is a common source of such contaminants. In several areas of Zacatecas state mining activities (some of them using cianuration) and tailing reprocessing activities are currently being developed. It was found that new mine tailings in the area are recklessly managed and there is an alarming lack of enforcement mechanisms to oblige the mining companies to obey the environmental laws and regulations. Those new tailings are undoubtedly a source of heavy metal contamination of the neighbouring agricultural land. This should be considered as a threat to health and food safety of the people in the region. Considering the high concentration levels found for arsenic, lead and mercury in soils of two irrigation zones of Guadalupe, mitigation activities should be implemented. Respiratory and ingestion routes are the most important sources of heavy metal exposure. There is an urgent need to conduct more research on potentially contaminated agricultural areas. Further health and environmental risk assessments should be promptly conducted in the region.

6. Aknowledgements

The authors wish to thank the Laboratory of Environmental Geochemistry of the Geology Institute at Universidad Nacional Autónoma de México (UNAM) for the facilities provided for this research. Osiel González Dávila wishes to thank the partial funding for this research provided by Consejo Nacional de Ciencia y Tecnología (CONACYT México) and Dirección General de Relaciones Internacionales – Secretaría de Educación Pública (DGRI – SEP México).

7. References

Allen, S.E., Grimshaw, H.M., Parkinson, H. M. & Quarmby, J.A. (1974). Chemical Analysis of Ecological Materials. Blackwell Scientific publications, Oxford.
Bidar, G., Garcon, G., Pruvot, C., Dewaele, D., Cazier ,F., Douay, F. & Shirali, P. (2007). Behaviour of Trifolium repens and Lolium perenne growing in a heavy metal

contaminated field: plant metal concentration and phytotoxicity. *Environmental Pollution*, 147, pp 546–553.

Black, G.R. (1965). Bulk Density: Method of Soil Analysis. Monograph No. 9 Part I. American Society of Agronomy Inc., Madison, WL.

Castro, A., Torres, L. & Iturbe, R. (2003). Risk of aquifers contamination in Zacatecas, Mexico due to mine tailings. In Tailings and Minewaste '03. Swets and Zeitlinger B.V., Lisse, The Netherlands. pp. 255 - 262.

Comisión Nacional del Agua (CONAGUA) (2010). *Estadísticas del Agua en México, edición 2010*. México D.F.

Food and Agriculture Organization of the United Nations (FAO) (2003) *Trade Reforms and Food Security: Conceptualizing the Linkages*. Rome, Commodity Policy and Projections Service, Commodities and Trade Division, FAO. 13.03.08, Available from:

ftp://ftp.fao.org/docrep/fao/005/y4671e/y4671e00.pdf

Figueroa, F., Castro–Larragoitia, J., Aragón, A. & García–Meza, J. (2010). Grass cover density and metal speciation in profiles of a tailings-pile from a mining zones in Zacatecas, North-Central Mexico. *Environmental Earth Sciences*. 60 (2), pp. 395-407.

González Dávila, O. (2011). Water Arsenic and Fluoride Contamination in Zacatecas Mexico: An Exploratory Study. 8th International Conference "Developments in Economic Theory and Policy." The University of the Basque Country, Spain. 05.08.2011, Available from:

http://www.conferencedevelopments.com/files/Gonzalez_Davila.pdf

Gutiérrez-Ruiz, M., Romero, F.M. & González-Hernández, G. (2007). Suelos y sedimentos afectados por la dispersión de jales inactivos de sulfuros metálicos en la zona minera de Santa Bárbara, Chihuahua, México. *Revista Mexicana de Ciencias Geológicas*, 24, pp. 170-184.

Horton, J. and Newson, D. (1953). A Rapid Gas Evolution For Calcium Carbonate Equivalent in Liming Materials. *Soil Sci. Soc. Am. Proc.* 17, pp. 414-415.

Instituto Nacional de Geografia e Informatica INEGI (2011). Banco de Información Económica (BIE). 01.08.2011, Available from: http://dgcnesyp.inegi.gob.mx/

Kabata-Pendias, A. (2001). Trace Elements in Soils and Plants third ed. CRC Press, Boca Raton, FL.

Leal, M. & Gelover, S. (2002). Evaluación de la calidad del agua subterranea de fuentes de abastecimiento en acuíferos prioritarios de la región Cuencas Centrales del Norte. Anuario IMTA 2002.

Manzanares, E., Vega, H., Letechipia, C., Guzmán, L., Hernández, V. & Salas, M. (2003) *Determinación de mercurio y plomo en la población la Zacatecana*. Universidad Autónoma de Zacatecas. México.

Marques, A.P.G.C., Moreira, H., Rangel, A.O.S.S. & Castro, P.L. (2009). Arsenic, lead and nickel accumulation in Rubus ulmifolius growing in contaminated soil in Portugal. *Journal of Hazardous Materials*, 165, pp. 174–179.

Melamed, D. (2004). Monitoring Arsenic in the Environment: A Review of Science and Technologies for Field Measurements and Sensors. Office of Solid Waste and Emergency Response, U.S. Environmental Protection Agency, Washington, DC, 26.09.2011, Available from:

http://www.epa.gov/tio/download/char/arsenic_paper.pdf

Méndez, M. & Armienta, M.A. (2003). Arsenic phase distribution in Zimapan mine tailings, Mexico. *Geofísica Internacional*, 42, pp. 131-140.

Mendoza-Amézquita, E.M., Armienta-Hernández, M.A., Ayora, C., Soler, A., Ramos-Ramírez, E. (2006). Potencial lixiviación de elementos traza en jales de las minas La Asunción y Las Torres, en el Distrito minero de Guanajuato, México. *Revista Mexicana de Ciencias Geológicas*, 23, pp. 75-83.

Moldovan, B.J., Jiang, D.T. & Hendry, M.J. (2003). Mineralogical characterization of arsenic in uranium mine tailings precipitated from iron-rich hydrometallurgical solutions. *Environmental Science Technology*, 37, pp. 837-879.

Nelson, D. W. & Sommers, L.E. (1982). Total carbon, organic carbon and organic matter. In Page, L. (Ed.), Methods of Soil Analysis. Part 2. Agronomy 9. American Society of Agronomy, Madison, WI, pp. 279-539.

Olsen S.R. & Sommers L.E. (1982). Phosphorus. In Page, A., Miller, R., Keeney, D. (Eds.) Methods of soil analysis, Part 2. Chemical and microbiological propierties. Madison, WI., pp. 403-427.

Ortega-Larrocea, M.P., Xoconostle-Cázares, B., Maldonado-Mendoza, I.E., Carrillo-González, R., Hernández-Hernández, J., Díaz-Garduño, M., López-Meyer, M., Gómez-Flores, L. & González-Chávez, M.C.A. (2009). Plant and fungal biodiversity from metal mine wastes under remediation at Zimapan, Hidalgo, Mexico. *Environmental Pollution*, 158, pp. 1922-1931.

Rizzi, L., Petruzzelli, G., Poggio, G. & Vigna Guidi, G. (2004). Soil physical changes and plant availability of Zn and Pb in a treatability test of phytostabilization. *Chemosphere*. 57(9), pp. 1039-1046.

Rotkittikhun, R., Kruatrachue, M., Chaiyarat, R., Ngernsansaruay, C., Pokethitiyook, P., Paijitprapaporn, A. & Baker, A.J.M. (2006). Uptake and accumulation of lead by plants from the Bo Ngam lead mine area in Thailand. *Environmental Pollution*. 144, pp.. 681-688.

Salas–Luévano, M., Manzanares–Acuña, E., Letechipía–de León, C. & Vega–Carrillo, H.R., (2009) Tolerant and Hyperaccumulators Autochthonous Plant Species from Mine Tailing Disposal Sites. *Asian Journal of Experimental Sciences*, 23(1), pp. 27-32

Santos–Santos, E., Yarto–Ramírez, M., Gavilán–García, I., Castro–Díaz, J., Gavilán–García, A., Rosiles, R., Suárez, S. & López–Villegas, T. (2006). Analysis of Arsenic, Lead and Mercury in Farming Areas with Mining Contaminated Soils at Zacatecas, Mexico. *Journal of the Mexican Chemical Society*. 50(2), pp. 57-63.

Secretaría de Medio Ambiente y Recursos Naturales SEMARNAT (2007). Norma Oficial Mexicana NOM-147-SEMARNAT/SSA1-2004, Que establece criterios para determinar las concentraciones de remediación de suelos contaminados por arsénico, bario, berilio, cadmio, cromo hexavalente, mercurio, níquel, plata, plomo, selenio, talio y/o vanadio. Diario Oficial de la Federación. Viernes 2 de Marzo de 2007.

Servicio de Información Agroalimentaria y Pesquera SIAP (2011). Reportes Estadísticos. 05.08.2011, Available from: http://www.siap.gob.mx/

Tu, S., Ma, L.A., MacDonald, G.E. & Bondada, B. (2003). Effects of arsenic species and phosphorus on arsenic absorption, arsenate reduction and thiol formation in excised parts of *Pteris vittata L. Environmental and Experimental Botany.* 51, pp. 121–131.

4

Sustainable Environment – Monitoring of Radionuclide and Heavy Metal Accumulation in Sediments, Algae and Biota in Black Sea Marine Ecosystems

Alexander Strezov
Institute for Nuclear Research & Nuclear Energy
Bulgaria

1. Introduction

The implementation of advanced methods and technology in the field of marine ecosystems studies is important to assess the impact of pollutants on marine ecosystems and biodiversity and biota interactions, including reliable study of radionuclide and heavy metal content in soils, sediments and algae. Creation of data bases for long term environmental management of pollution; evaluation of ecological impact of human activities and management of sustainable environment – coastal zone and continental shelf, including assessment and forecasting techniques, is needed to understand the impacts of various activities on ecosystems, to contribute to protect the ecosystems from pollution and to develop long-term models for management of the coastal zone.

Improved knowledge of marine processes, ecosystems and interactions will facilitate the basis of new ecological status, the sustainable use of the marine environment and resources, while fully respecting ecosystem integrity and functioning, and to promote the development of new, integrated management concepts.

That is why the determining of variations of ecosystem functioning; concepts for a safe and environmentally responsible use of the seafloor and sub-seafloor resources; management models of transport pathways and impacts of pollutants, key elements and nutrients in marine environments is the main task of modern marine radioecology.

Another target is to develop a predictive capability for variations in ecosystem functioning and structure for better assessment of naturally occurring mechanisms of ecosystem behaviour. Research activities will address the effects of environmental factors and interactions at sea boundaries and interfaces in relation to ecosystem functioning and alteration, distinguish natural from anthropogenic variability, assess the role of extreme environments and their communities. The obtained results will facilitate the assessment of sedimentary systems for the sustainable management and use of the sea shelf.

Transport pathways and impacts of pollutants, key elements and nutrients in the marine environment will support the implementation and further development promote the advancement of relevant conventions for the reduction of nutrient and pollutant loads of the sea. Research activities should address the transport, cycling, coupling and accumulation of nuclide and heavy metal pollutants. The impact of pollutants should be addressed, their uptake by organisms, their ecotoxicological effects and the synergistic effects of multiple pollutants upon the marine ecosystems.

Another important task is the reducing the anthropogenic impact on biodiversity and the sustainable functioning of marine ecosystems, facilitating the development of safe, economic and sustainable exploitation technologies requiring safe and economic (yet sustainable) exploitation of marine resources by improving the scientific basis of sustainability.

Complex monitoring of Global changes and marine ecosystems for modelling and adequate management for sustainable environment have to include the foremost requirement to better characterise, observe and monitor the marine environment.

The objective is to stimulate "clean" technological developments for oceanic environments, to develop integrated coastal management concepts, including cost-benefit assessments, to alleviate pollution, flooding and erosion, in particular of fragile coastlines, and to ensure sustainable resource utilisation. Anticipated deliverables are integrated management tools and concepts for the coastal zone ecosystem; long-term predictions of coastal zone changes; reliable, economic and environmentally compatible coastal protection measures against flooding and erosion; effective monitoring in coastal, shelf and slope areas.

Emphasis should be put on efficiency, speed, reliability, environmental friendliness and safety. In parallel, attention will be given to the improvement of sample collection and handling (including samples from sea drilling), the exploitation of distributed sample collections, and inter-calibration exercises, networking and joint experiments. Special attention will be given to the problems of coastal inlets (tidal basins, estuaries, lagoons, arias, brachial zones).

2. Radionuclides in the marine environment

Since 1945 there has been a continuous release of technogenic radionuclides to the environment, leading to their accumulation in the seas and oceans, the main source of contamination being nuclear weapons tests in the atmosphere, on land and in water. The Danube was the main contributor for nuclides in the Black Sea region before 1986 (mainly ^{90}Sr and ^{137}Cs) (Polikarpov 1966; Kulebakina et al., 1984; Sokolova et al., 1971), while the Dnyepr took second place.

After the Chernobyl accident a considerable increase occurred in nuclides quantity in the environment (Buesseler K., 1987; Kulebakina et al, 1988, 1989; Baumann, et al, 1989; Broberg 1989; Hadderingh et al, 1989). The Black Sea being near to the reactor accepted great quantity direct atmospheric radionuclide fallout. Additionally Chernobyl nuclides entered and continued to enter the environment carried by the big rivers - Dnyepr, Dnester, Danube.

A major part of the distribution and migration of long lived nuclide contaminants (mainly ^{137}Cs, ^{90}Sr) are not only the hydrophysical processes but also the biosedimentation and sorption on bottom sediments and concentration by biota. It is known that ^{137}Cs and ^{90}Sr fallout before 1986 are found in sea water mainly in soluble forms (Shvedov et al., 1962), and are weakly affected by biological migration (contrary to other nuclides as ^{141}Ce, ^{144}Ce, ^{103}Ru,

Sustainable Environment – Monitoring of Radionuclide and Heavy Metal Accumulation in Sediments, Algae and Biota in Black Sea Marine Ecosystems

53

^{95}Zr etc.). Chernobyl radionuclides are characterized by a higher content of nonsoluble forms (even "hot" particles) which are part of the reactor fuel. These peculiarities affect the behavior of the Chernobyl nuclides in sediments and their accumulation in biota. The contamination of sea bed sediments by ^{137}Cs and ^{90}Sr had a "spot" character which was due to the way Cs entered the environment at the beginning - mainly airborne on different particles.

Cesium penetrated rather quickly the water depths - in 1987 Cs was measured up to 120 m, while in 1988 it reached 200 - 250 m. The observed phenomenon of "self decontamination" of Black Sea waters was due to the radioactivity spots dilution and secondly to interaction of dissolved Cs with the bottom sediments (Polikarpov 1987, 1988, Kulebakina 1988, 1989).

The natural nuclides members of uranium 238 and thorium 232 also enter the environment be several pathways and are also accumulated in biota.

So it is necessary to assess scientifically the impacts of discharges on humans and to identify the most important critical 'pathways' or 'groups' and the risks involved also attempt to consider multiple exposure routes. It can indeed be argued that radioactive pollution control is more successful in producing a mature methodology which is unifying in respect of all radioactive substances and there are only a few cases where conventional pollutants are successfully ranked against each other - such as in considering their relative importance as 'greenhouse' gases or the relative toxicity of certain similar organic compounds. However, in conventional pollution control, one further step is taken after risk assessment, one which is absent from radioactive pollution control - namely, the information generated in the earlier stages is used to produce 'environmental quality standards' against which absolute concentrations, trends and the effectiveness of pollution control at source can be clearly measured. Thus, if instead of considering, e.g. plutonium or technetium in units of 'concentration', we consider them as contributors to the overall 'dose', we cease to treat them as 'substances' in their own right and simultaneously regard the world as a 'brown-field site' already 'contaminated'. However, ecological toxicology (ecotoxicology) is required for predicting real world effects and for site-specific assessments. Ecotoxicology and ecology have shown similar developmental patterns over time; closer cooperation between ecologists and toxicologists would benefit both disciplines. Ecology can be incorporated into toxicology either extrinsically (separately, e.g., providing information on pre-selected test species) or intrinsically (e.g., as part of test species selection) - the latter is preferable. General guidelines for acute and chronic testing and criteria for species selection differ for ecotoxicology and environmental toxicology, and are outlined. An overall framework is proposed based on ecological risk assessment, for combining ecology and toxicology (environmental and ecological) for decision-making.

3. The Black Sea marine ecosystem

The pollution of marine ecosystems by nuclides and heavy metals has been a world-wide problem in the last decades. The Black Sea ecosystem and ecological status has been damaged mainly as a result of chemical pollution. Much of the pollutants come from major rivers and from smaller sources in all Black Sea coastal countries. The Black Sea water column is heavily impacted by a great number of pollutants originating from different sources of direct and indirect discharge of land based sources - sewage, fallout etc. from economies of coastal states.

The Black sea is a half-enclosed sea, with 40°27'N-46°32'N latitude and 27°27'E- 41°42'E longitude. Together with Azov Sea, it covers an area of 462000 km². Its east to west dimension is 1150 km and from north to south is 610 km. In the Black sea main basin, the depth of water approaches 2200 m, and the western shelf zone is comparatively shallow. The Black sea is surrounded by six countries and is linked with the Mediterranean sea through the Bosphorus, and the Azov Sea to the north.

The Black Sea has experienced the worst environmental degradation of all of the world's oceans. The situation has become so severe that it has affected the health, well being, and standard of living of the people in the immediate area. Most of the six coastal countries - Bulgaria, Georgia, Romania, Russia, Turkey, and Ukraine - have unstable or collapsed economies. The Black Sea's area is 431,200 km². About 160 million people live in the Black Sea catchment basin, including 80 million only in the Danube River basin. Although international agreements, strategic plans, and national environmental programs are in place, the severe economical problems have significantly slowed environmental monitoring, remediation, and restoration efforts. The environmental crisis and subsequent dramatic changes in the Black Sea's ecosystem and resources are a direct effect of both natural and anthropogenic causes: an enormous increase in the nutrient and pollutant load from three major rivers, the Danube, Dnyestr, and Dnyepr; from industrial and municipal wastewater pollution sources along the coast; and from dumping on the open sea. The coastal industries discharge wastes directly into the sea with little or no treatment. The countries of the Black sea basin do their efforts to protect the nature of the sea by formulating international rules for the cleaning of water areas from oil and wastes. Nuclides and heavy metals in the marine environment constitute a potential risk to the flora and fauna species, including humans through food chains. Furthermore, there is increasing evidence that presence of nuclides and heavy metals is linked to the exacerbation of some microbial diseases in aquatic organisms. At sufficiently high concentrations, heavy metals are toxic to the organism, and so, it is important their concentrations to be monitored especially when increased above the normal levels in the environment before the effects on marine organisms. Nuclides and heavy metals are the most harmful elemental pollutants and are of particular concern because of their toxicities to humans. They include both essential elements like iron and toxic metals like cadmium and mercury. Most of them show significant affinity to sulphur and disrupt enzyme function by forming bonds with sulphur groups in enzymes. Cadmium, copper, lead and mercury ions bind to the cell membranes hindering the transport processes through the cell wall. Some of the metalloids, elements on the borderline between metals and nonmetals, are significant water pollutants (Manahan, 1999).

The concentrations of rivers pollutants (domestic and industrial discharges) provide useful information about the sources that have the potential to lead to local pollution problems along the Black sea coast. The shallow, biologically productive layer of the Black sea receives water from a waste drainage basin of about 17 countries. Pollutants, transported by the rivers, constitute the main source of pollution in the Black sea.

According to Zaitzev, (1992), the estimation of land based sources, the sea annually receives big amounts of mineral nitrogen, mineral and organic phosphorus, as well oil and oil products, detergents, zinc, lead, mercury, copper, arsenic, chromium.

Aquatic organisms, especially macroalgae, are widely used as bioindicators for the study of marine contamination by radionuclides and heavy metals. Some species tolerate high

Sustainable Environment – Monitoring of Radionuclide and Heavy Metal Accumulation in Sediments, Algae
and Biota in Black Sea Marine Ecosystems

55

levels of pollutants and can successfully be used to obtain reliable information of marine ecological status.

4. Radionuclides in Black Sea marine ecosystems

Radionuclides are a part of anthropogenic pollutants in the Black sea marine ecosystems. Massive amounts of industrial effluents are transported by the big rivers that enter the Black sea (Danube, Dnyeper, Dnester etc). The change in the radiation situation in the Black Sea after the Chernobyl accident stimulated multiple studies of radionuclide accumulation processes in biota as the Black sea received a great amount of radionuclides, due to its geographical position. The complex analysis of pollutants is a major task for modern ecology in obtaining reliable information about the type and quantities of substances entering the marine environment. The analysis of environmental matrixes, such as water/sediments/algae, provides a picture of the total contaminant load in a given ecosystem.

The contamination of Black sea littoral zone is a powerful factor affecting the phytobentos dynamics. The technogenic and natural nuclide releases due to human activities in the marine ecosystems lead to a change in contaminant content and may affect the composition of species in the marine environment (Bologa et al 1996, Guven et al. 1993). The complex analysis of pollutant concentrations in the marine environment gives reliable information for the types and quantities of contaminants that enter the hydrosphere. The change in the radiation situation of the Black sea after the Chernobyl accident was the reason for multiple studies on radionuclide accumulation processes in biota. The Black sea received a great amount of radionuclide impact due to its geographical position as the closest marine basin to the accident site. The technogenic radionuclides got into the marine ecosystems through atmospheric fallout and through the big rivers Danube, Dnieper and Dnester that enter the northwest Black sea corner. For this reason, the level of anthropogenic nuclides should be monitored to evaluate the radioactivity transfer along the trophic chain and assess the radiation risk for biota in the marine ecosystem.

Since 1945 there has been a continuous release of technogenic radionuclides to the environment started, leading to their accumulation in the seas and oceans, the main source of contamination being nuclear weapons tests in the atmosphere, on land and in water. The Bulgarian Black Sea coast received a great amount of radionuclide impact during nuclear tests in 1960[ies] and after 1986 being close to Chernobyl NPP, a considerable direct atmospheric radionuclide fallout occurred in the environment, Additionally, Chernobyl nuclides entered the marine environment mostly in the northwest corner of the Black Sea carried by the big rivers - Dnyepr, Dnester, Danube (Keondjan et al., 1990) plus pollutant emissions carried by the local rivers (Tuncer, *et al.*, 1998).

Marine sediments are widely used for environmental control because of their ability to accumulate various pollutants. Macroalgae are another important medium for nuclide accumulation in marine ecosystems. Radionuclides affect the living organisms both as heavy metals and by their radiation. They participate in radionuclide and heavy metal transfer to the biosphere and man as elements of the food chain of marine biota. Many authors have investigated migration of radionuclides as well as biological effects of ionizing radiation in the Black sea environment.

A monitoring program for measuring technogenic and natural radionuclides in marine environmental samples from the Bulgarian Black sea coast has been utilized since 1991.

The radionuclide content dependency of on the season, location, type of sediment and type of algae and the comparison of radionuclide content in bottom sediments and algae from one and the same sampling location gives information for the mechanisms of radionuclide transfer from the sediments to biota as well as the trend of the potential hazard for the marine ecosystems.

5. Radionuclides in the Black Sea sediments

A major part of the distribution and migration of long lived nuclide contaminants (mainly ^{137}Cs, ^{90}Sr) are not only the hydrophysical processes but also the biosedimentation and sorption on bottom sediments and concentration by biota. It is known that ^{137}Cs and ^{90}Sr fallout are found in sea water mainly in soluble forms and are weakly affected by biological migration (contrary to other nuclides as ^{141}Ce, ^{144}Ce, ^{103}Ru, ^{95}Zr etc.). Chernobyl radionuclides are characterised by a higher content of nonsoluble forms (even "hot" particles) which are part of the reactor fuel. These peculiarities affect the behaviour of the Chernobyl nuclides in sediments and in their accumulation in biota. The contamination of sea bed sediments by ^{137}Cs and ^{90}Sr had a "spot" character which was due to the way Cs entered the environment at the beginning - mainly airborne on different particles.

Caesium penetrated rather quickly the water depths - in 1987 Cs was measured up to 120 m, while in 1988 it reached 200 - 250 m. The observed phenomenon of "self decontamination" of Black Sea waters was due to the radioactivity spots dilution and secondly to interaction of dissolved Cs with the bottom sediments (Polikarpov et al 1987, Kulebakina et al 1988, 1989).

The association of radionuclides with sediments in coastal and estuary areas makes the sediments a large reservoir for radionuclides affecting significantly specific marine ecosystems. As the sea water is constantly in contact with organic and inorganic matter from sea bed sediments, the first necessary step in the estimation of radionuclide migration is the measurement of radionuclide content in sea bed sediments (as a first stage in food chain towards man) and set up a data base for radioisotope content in the Black Sea.

The measurement of technogenic and natural radionuclides in sea bed sediments was carried out in each season of every year after 1991 - spring, summer and autumn, because in this way the estimation of concentration variations of the contaminants, their accumulation and influence on marine ecosystems can be followed. The intercomparisson between the data for the technogenic and natural radionuclides gives the whole picture of isotope impact at a chosen reference point.

Many authors - Russian and Ukrainian scientists, (Polikarpov, Kulebakina et al., 1988, 1989,), Turkish – Topcuoglu 2001, Guven 2002, Bulgarian (Strezov 1996, 1998; Romanian – Bologa, Patrascu 1996) have performed measurements in the Black Sea.

The impact of Chernobyl on marine ecosystems was intensively studied and much data were published in recent years. Many authors investigated the distribution of Chernobyl radionuclides and their accumulation in lake sediments as well as in sea bed sediments.

The coast of Bulgaria, extending 270 km is mainly soft sedimentary limestone or sandstone, overlaid in many areas with beach or wind-blown sand. In the north, suspended material from the Danube delta has settled, carrying a lot of intertidal mud. The association of

Sustainable Environment – Monitoring of Radionuclide and Heavy Metal Accumulation in Sediments, Algae
and Biota in Black Sea Marine Ecosystems

57

radionuclides with sediments in coastal and estuary areas makes sediments a large reservoir for radionuclides affecting significantly specific marine ecosystems. As the seawater is constantly in contact with organic and inorganic matter from sea shelf, the first necessary step in the estimation of radionuclide migration is the measurement of radionuclide content in sea bed sediments and setting up a radioecological data base. Receiving relevant information for radionuclide concentrations in coastal waters and sediments is an important stage in realization of monitoring and control of marine ecosystems at the coast (Fig 1).

The main method for measuring the nuclei content in sediments and algae was the high resolution gamma spectroscopy performed on large high purity Ge semiconductor detectors with nuclear electronic tracts using sophisticated software.

The upper layer of sediments was collected from approximately 1 m^2 of the bottom acquiring about 2-3 kg of solid phase plus several litres of aqueous phase. The depth of the sample layer was maximal 3 cm to evaluate radionuclide content on the surface of the sea bed. In this way the potential seasonal variations could be estimated. The sample collection was performed by experienced scuba divers who carefully selected the sample site to avoid or minimize the differences of samples in particle size, depth and distance from shore, so that the comparison of the sea bed samples of different seasons could be done with greater level of confidence, reliability and accuracy.

Samples were also collected and data obtained for the main Black Sea resorts - Albena, Golden sands, Sunny Beach etc. as well as for some of the main cities along the Bulgarian Black Sea coast. These data were compared with samples taken from definitely clean areas. The contribution of the inflowing rivers for radioisotope content in the Black Sea was also studied sampling river's estuaries and comparing them with the inland sections of the same rivers. So the influence of the river on the adjacent area was compared with that of the sea.

The obtained results (Fig. 2) for Black Sea sediment samples show that radionuclide concentrations strongly depend on the nature of the sea bed sediments, because the data obtained for sand sediments are within a close range while those for silt and slime ones are higher and vary to a much greater extent.

The beach matrix from the near shore sediments at these locations is mainly sand and ^{137}Cs data are within a close range: Sunny Beach - 3.2 - 5.6 Bq/kg, Golden Sands – 1.8 – 6.9 Bq/kg, Albena - 3.4 – 7.3 Bq/kg, Tulenovo 4.0-7.1 Bq/kg, Kamen Briag - 4.4-6.6 Bq/kg, Balchik - 4.6 - 7.8 Bq/kg, Primorsko 4.0 - 5.6 Bq/kg, Sinemoretz – 3.6 – 7.8 Bq/kg. It should be noted that all sand sediment data fall within 8 Bq/kg level except Albena, Golden sands, Ravda, Burgas and Sozopol where nuclide content is higher.

The highest measured cesium content (Fig. 3) on the Bulgarian Black Sea coast is at the north locations with slime sediments – Kaliakra (mean 89 Bq/kg), Kavarna (mean 30 Bq/kg) and central Ravda2 (mean 65 Bq/kg). This fact can be attributed to the influence of the big rivers Danube, Dnyepr, Dnester, entering the northwest part of the Black Sea.

The increase in ^{137}Cs concentration in slime sediments and sorption on fine particles leads to caesium scavenging and occurrence at greater depths, which is due to physico-chemical interaction processes of the soluble Cs forms with the surrounding media. In sand and sandy sediments Cs content does not change greatly while the process of ^{137}Cs accumulation is observed in slime and silt sediments. Due to such a process, sea bottom sediments play a major role in radionuclide redistribution between different components in the ecosystems, which change the concentration of ^{137}Cs in the water as it is accumulated more in the sediments.

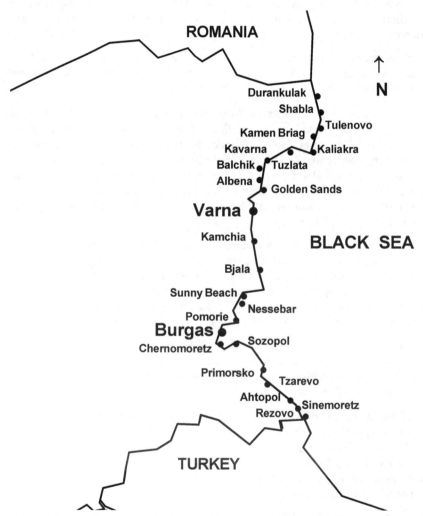

Fig. 1. Scheme of sampling locations along the Bulgarian Black Sea coast

The observed dependence of radionuclide content on sediment type is valid also for the natural nuclides in sediments. The lowest concentrations of natural nuclides is in the sand sediments from the north locations – Duran Kulak, Shabla, Tulenovo, Kamen briag and the measured natural nuclide concentrations are: ^{238}U (4.0 - 10) Bq/kg, ^{232}Th (3.4 - 10) Bq/kg, ^{226}Ra (3.6 - 9) Bq/kg. Low content was obtained at Sunny beach, Nessebar and Primorsko: ^{238}U (3 – 8) Bq/kg, ^{232}Th (4.4 - 8) Bq/kg, ^{226}Ra (3.5 – 6.2) Bq/kg (Fig 2).

Sustainable Environment – Monitoring of Radionuclide and Heavy Metal Accumulation in Sediments, Algae and Biota in Black Sea Marine Ecosystems

59

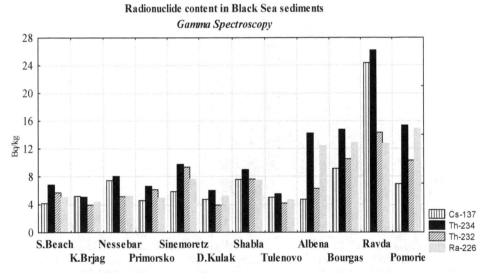

Fig. 2. Nuclide content in sand sediments along the Bulgarian coast

The obtained mean values for silt sediments are between sand and slime ones with exception of Bjala, whose values for ^{238}U are in the range14 - 77 Bq/kg; ^{232}Th 12- 110 Bq/kg; ^{226}Ra 10 - 77 Bq/kg and are the highest measured at the silt and slime locations (Fig.4, 5). The data on Fig. 3 show that all natural nuclide content in slime sediments varies around 30 Bq/kg, (except ^{232}Th at Kavarna, ^{238}U at Maslen nos and Chernomoretz), showing some uniformity of natural nuclide concentrations along the whole coast.

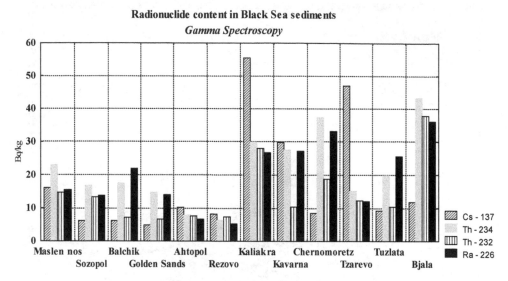

Fig. 3. Mean nuclide content in silt and slime sediments along the Bulgarian coast

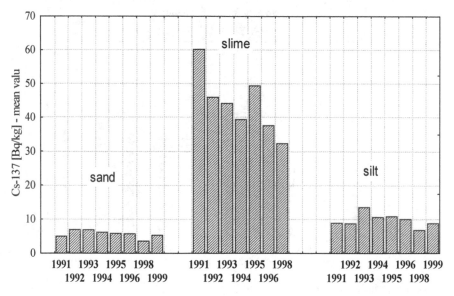

Fig. 4. Mean ^{137}Cs content in sediments during the period 1991 - 1999

The highest values for natural nuclides content (similarly to Cs) are obtained for the slime sediments - the obtained results for ^{238}U vary in the different years in the range 5 - 50 Bq/kg, ^{232}Th - 4.0 - 35 Bq/kg and ^{226}Ra - 9 - 50 Bq/kg. The mean values of ^{238}U, ^{232}Th and ^{226}Ra specific activities for slime sampling locations are presented on Fig. 3 and these values show the maximum natural nuclide content at the Bulgarian Black Sea coast. The obtained

results show that there is a similarity between the accumulation of ^{238}U and ^{232}Th in Black
Sea bed sediments. The measured U and Th values are within the range of the cited in the
literature meaning that there is no serious contamination with U and Th at the Black Sea
coast. ^{226}Ra content generally follows the pattern of U and Th with few exceptions.

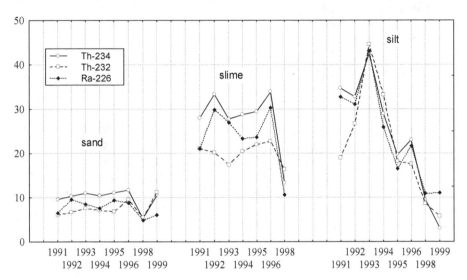

Fig. 5. Natural nuclide content (Bq/kg) in Black sea sediments in the 1991 – 1999 period

Samples were taken from deep sediments (56 – 155 m) from the Bulgarian territorial waters
plus two samples at maximal depth 2040 m and also were measured for radionuclide
content.

The obtained results (Fig. 6) show that Cs content is rather low (compared to the shelf),
while natural nuclide concentrations increase with the depth. ^{226}Ra nuclide concentration is
higher in the middle depths (55 – 90 Bq/kg) than other natural nuclides while at the bottom
2000 m U content is the highest (90 – 135 Bq/kg). The character of deep sediment samples is
slime except at 2000 m where the matrix is very hard in structure and black in color.

The multivariable (cluster) analysis of all measured sediment samples (145) for eight
consecutive years depending on all measured nuclides shows that the type of sediment is a
basic factor for nuclide accumulation in sediments (Fig. 6). The nuclide values for all sand
sediments are combined in one cluster (from Albena to Shabla max Eucledian distance is
2.5). The second cluster includes locations close in geographical position and sediment type
(slime) - Tuzla, Kavarna and Tzarevo, while Bjala, Chernomoretz and Kaliakra are
completely separate from the rest.

The performed correlation analysis for tree consecutive season, for each type of sediment at
all locations shows, that there is no clear season dependence of isotope concentrations - the
calculated correlation coefficients between different seasons are close to 1. The statistical
analysis of all nuclide data for all sampling sites (Fig 7) groups together sites with similar
sediment matrix (sand, silt or slime) which supports the assumption of nuclide sorption
dependence on the type of sediment matrix.

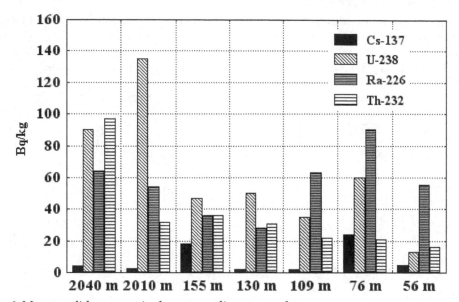

Fig. 6. Mean nuclide content in deep-sea sediment samples

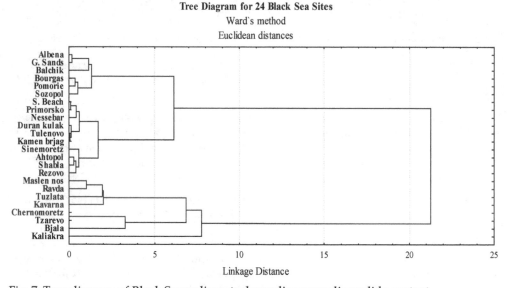

Fig. 7. Tree diagram of Black Sea sediments depending on radionuclide content.

6. Nuclide content in macroalgae

The data on the radionuclide content in different algae species are presented in Table 1 and in Figs. 5 and 6. The ^{137}Cs content in different algae species vary on average between 3 and 20 Bq/kg. The content of natural nuclides is close to the lowest limits of detection (LLD).

Sustainable Environment – Monitoring of Radionuclide and Heavy Metal Accumulation in Sediments, Algae
and Biota in Black Sea Marine Ecosystems

63

The most interesting species is *Bryopsis plumosa* whose ^{226}Ra and ^{210}Pb nuclide contents are with some orders of magnitude higher than the same nuclides in other species.

The ^{226}Ra contents in Black sea macrophytes vary in the range 2-25 Bq/kg for brown and 3-18 Bq/kg for green species. The accumulation intervals for ^{226}Ra are close to each other for the two brown species and for *Ulva*, *Cladophora* and *Enteromorpha*, respectively.

The mean values of Ra vary for different species in the range 6.8-11.7 Bq/kg and can be arranged as follows:

Ceramium rubrum > *Clad. vagabunda* = *Cyst. barbata* > *Cyst. crinita* >
Calith. corumbosum = *Ulva rigida* = *Ent. intestinalis* > *Chaetom. Gracilis.*

It is evident that the mean values are close to each other and if they are combined, the total mean for all brown and green algae is the following: 10+1 Bq/kg (N = 40, range 2-25 Bq/kg) for brown species and 8.6+0.7 Bq/kg (N = 40, range 3-28 Bq/kg) for the green ones. Mean Ra content in red algae is 15 Bq/kg, range 2-39 Bq/kg, which shows similar pattern of natural nuclide accumulation by the algae species along the Black sea coast. ^{210}Pb content in all algae species is in the range 3 – 40 Bq/kg, which is close to the values of neighbouring Mediterranean Sea and shows the level of this nuclide from recently deposited particles.

Nuclide	No of samples	Mean value ± SD	Minimum value	Maximum value
Chlorophyta				
^{40}K	40	1009 ± 90	300	2500
^{137}Cs	40	4.2 ± 0.5	1.2	13.0
^{226}Ra	40	10 ± 1	3	28
^{210}Pb	40	8.8 ± 0.9	3	22
Phaeophyta				
^{40}K	40	1492 ± 53	1025	2250
^{137}Cs	40	6.1 ± 0.4	2.6	12.0
^{226}Ra	40	10 ± 1	2	25
^{210}Pb	40	12 ± 1	2	30
Rhodophyta				
^{40}K	36	1146 ± 110	90	2100
^{137}Cs	36	7.4 ± 0.8	1.2	18.0
^{226}Ra	36	15 ± 2	3	39
^{210}Pb	36	15.8 ± 1.5	4	40

Table 1. Nuclide content (Bg/kg) in Black sea algae

The radionuclide content in marine ecosystems and especially in algae is in the limits of published data for the Black Sea and the Mediterranean.

The ^{137}Cs, ^{226}Ra and ^{210}Pb content of the radionuclides has been measured in sediments and macroalgae located in several geographic zones along the Bulgarian Black sea coast, during the period from 1991 to 1999. The accumulation capacity and radionuclide content for different algae species also was determined for three algae phyla in the marine environment. No great difference was found in nuclide accumulation in algae, but the red species seem to accumulate nuclides to a higher degree than the brown and green species.

The data show that macrophytes can be used as reliable indicators for marine environmental assessment. With this paper we intend to fill the lack of data concerning radionuclides in sediments and macroalgae along the whole Bulgarian Black Sea coast. The data can be used as reference levels for further monitoring and control of the marine ecosystem status.

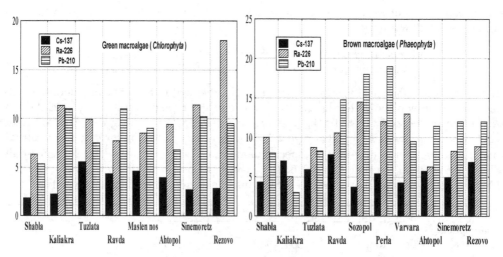

Fig. 8. Nuclide contents (Bq/kg) in green and brown algae

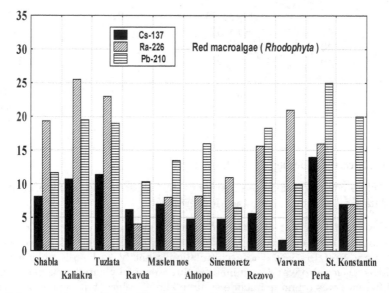

Fig. 9. Nuclide contents (Bq/kg) in Black Sea macrophytes - *Rhodophyta*

Sustainable Environment – Monitoring of Radionuclide and Heavy Metal Accumulation in Sediments, Algae and Biota in Black Sea Marine Ecosystems

65

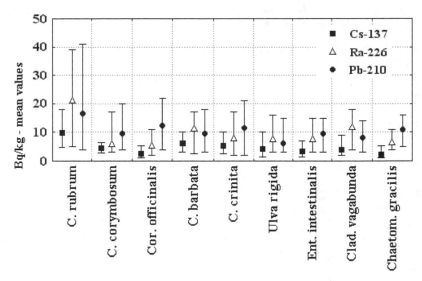

Fig. 10. Comparison of nuclide contents (Bq/kg) in Black Sea algae in different algae species

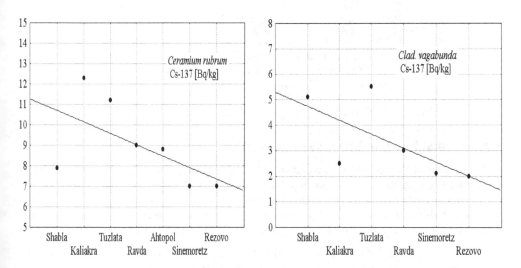

Fig. 11. ^{137}Cs content in macroalgae – local variations

The analysis of all alga samples for natural radionuclides content from ^{238}U series - ^{226}Ra and ^{210}Pb. Showed ^{226}Ra concentration values in the range 2 – 28 Bq/kg for red (mean 15 ± 2):, for brown 2 - 17 Bq/kg (mean 9.8 ± 0.8) and for green 3 - 18 Bq/kg (mean 9 ± 1). Accumulation intervals for ^{226}Ra are close for the two brown and for green *Ulva rigida, Clad. vagabunda* and *Ent. intestinalis*. The ^{226}Ra mean values in different species can be arranged as follows:

Cer. rubrum > *Clad. vagabunda* = *Cyst. barbata* > *Cyst. crinita* = *Ulva rigida* = *Ent. intestinalis* > *Chaetom. gracilis* > *Callith. corymbosum* = *Cor. officinalis.*

The obtained ^{210}Pb average values content in studied algae vary between 4 and 35 Bq/kg (mean 15 ± 1) for red, 2 and 21 Bq/kg (mean 11 ± 1) for brown and from 3 to 16 Bq/kg (mean 8 ± 1) for green algae. If ^{210}Pb mean values are arranged depending on algae species, the following order is obtained:

Cer. rubrum > *Calith. corymbosum* > *Cor. officinalis* = *Cyst. crinita* > *Chaetom. gracilis* > *Ent. intestinalis* > *Cyst. barbata* > *Clad. vagabunda* > *Ulva rigida*

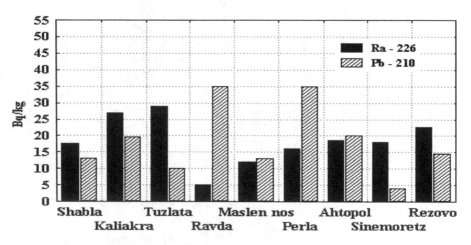

Fig. 12. Mean ^{226}Ra and ^{210}Pb content (Bq/kg) in *Ceramium rubrum*

7. Heavy metal levels in Black Sea sediments

The distribution of the recent bottom sediments in the Black sea shows a variable pattern. The sediment composition and origin depend on the provenance areas, hydrodynamic and lithodynamic activity in the contact zone of the sea, and on the morphology of the bottom topography. Sediment formation is also influenced by the solid riverine runoff and coastal abrasion, slope-derived supply, and biogenic and chemogenic matter (Ignatov, 2008).

Heavy metal concentrations in surface sediments can provide historical information on heavy metal inputs at that location. Such surface sediment samples are also used as environmental indicators to reflect the current quality of marine systems for many pollutants. (Förstner et al., 1980). Nijenhuis et al., (1999) reported that the enrichment of trace elements in marine sediments may, in general, originate from the following sources; super and subjacent sediments, through diagenesis; suboxic shelf and slope sediments, hydrothermal input; aeolian input; fluvial runoff; sea water. Heavy metal levels in sediment from the Black Sea were investigated by many researchers (Ergul et al., 2008; Topcuoglu et al., 2002; and 2004, Strezov et al., 2003) that measured heavy metal concentrations in sediment samples and found dependence upon seasonal changes within the water column as well as anthropogenic and geological inputs such as weathering and run-off from land-based sources. Furthermore, metal concentrations in both surface sediments and sinking particles also suggest that heavy metal concentration is generally enhanced in the eastern region along the Black sea coast of Turkey.

Bulk heavy metal (Fe, Mn, Co, Cr, Ni, Cu, Zn and Pb) distributions and their chemical partitioning together with total organic carbon and carbonate data were studied in 0-2 cm oxic to anoxic surface sediments, obtained at 18 stations throughout the Black sea by Kiratli et al., (1996). Chemical partitioning of the heavy metals revealed that Cu, Cr and Fe seem to be significantly bound to the detrital phases whereas carbonate phases tend to hold considerable amounts of Mn and Pb.

Coban et al., (2009) found heavy metals in sediment at significant levels of 0.47 µg/g for Cd, 67.95 µg/g for Cr, 30.21 µg/g for Cu, 274.4 µg/g for Mn, 37.03 for µg/g Ni, 39.14 µg/g for Pb and 84.6 µg/g for Zn, that were comparable with those found in the estuarine areas of other countries in the region.

Topcuoglu et al., (2003b) determined the radionuclides ^{137}Cs, ^{238}U, ^{232}Th and ^{40}K and Cd, Pb, Cu, Zn and Mn in sediment samples collected from two stations at the eastern Turkish coast of the Black sea. The result from this study showed that radionuclide concentrations in the sediment fraction were significantly higher because of the influence of collection sites. In general, the heavy metal concentrations in that study were not higher than those previously observed. However, Pb and Cu levels increased in sediment in the Turkish area of the Black Sea during the investigated years.

In conclusion, heavy metal pollution in the Black sea has attracted considerable research attention since last 20 years. Sources of heavy metals in the Black sea ecosystem can be mainly attributed to terrestrially derived waste water discharges, agricultural and industrial run-off, river run-off atmospheric deposition of combustion residues, and shipping activities. It is clear from many studies conducted that the heavy metal pollution in the Black sea should be taken into account. In the last 20 years, in some areas of Black sea, metal concentrations in sea water exceeded the accepted levels. Especially, lead and cadmium levels were found higher in fish species than the legal limit for human consumption. High levels of heavy metals have been reported by the authors, suggesting that heavy metal pollution in algae and sediments from certain regions of Black Sea is rather high.

The Black Sea is a unique ecosystem because it is an inner sea with low salinity, half-isolated from the Mediterranean with hydrology and phytobentos different from the other seas in the same biogeographic region.

The antropogenic contamination of marine ecosystems is a very important stress factor and defines the necessity for systematic monitoring and control of contaminants (heavy metals – HM, radionuclides, etc.) that affect marine biota. The main sources of Black Sea pollution are atmospheric fallout, the big rivers run-off as well as local pollutant emissions (Tuncer et al., 1998).

Macrophytic algae, being one of the primary stages in the trophic chain, play a major role in marine ecosystems (Kilgore et al., 1993). Algae interact with the environment through processes that include chemical bioconcentration, excretion, organic matter production and decomposition (Carpenter and Lodge, 1986). They have been used as a signal for the living status of marine ecosystems and considered as valuable indicators for HM assessment in the major components of the water ecosystems because of their accumulation capacity (Forsberg et al., 1988). Some algae possess ecological mutability so they can survive in contaminated habitat (Kalugina-Gutnik, 1975). Closely related species may exhibit different accumulation capabilities for trace elements, so there is a need to identify indicators that are biologically dominant and widespread in the ecosystems. This will allow intraspecific comparison of accumulated metal concentrations over large geographical areas (Rainbow, 1995; Rainbow and Phillips, 1993).

HM are among the most studied contaminants in marine ecosystems. Their effect is through direct poisoning as well as by accumulation and transfer along the trophic chain, by which they influence the functioning of biosphere (Babich et al., 1985).

Metal levels in algae reflect local geology or local anthropogenic activities and the contamination is generally similar to background levels of the sites. Today's HM concentrations in marine environment are generally more than 10 times higher than it was in prehistoric times. The levels are consistently higher in surface waters than in deeper layers. The macrophytic species from phylum Chlorophyta (green) are widely distributed in the coastal Black Sea ecosystems and some of them (*Ulva rigida, Enteromorpha intestinalis, Cladophora vagabunda*) can be found in almost all areas and environmental pollution was studied by means of green algae. These species have been extensively used to monitor marine pollution for Mn, Cu, Pb and Cd in various geographical areas (Favero et al., 1996; Muse et al., 1999).

Many authors have investigated biogeochemical migration of anthropogenic contaminants (HM, radionuclides) in the Black Sea environment (Polikarpov et al., 1991; Roeva, 1996; Guven et al., 1992; Topcuoglu et al., 1998, 1999, 2001) while information about heavy and toxic metals at the Bulgarian coastal zone is scarce and insufficient.

Data were obtained by Atomic Absorption (AAS) and X-Ray fluorescence analysis for the accumulation and seasonal distribution of Fe, Mn, Cu, Pb and Cd in six green macrophytes, collected from eight locations along the Bulgarian Black Sea coast during the period spring 1996 to summer 2002. Fe has a great binding capability for alga lipids and is accumulated to the highest degree in the Black Sea green macrophytes.

The measured Fe concentrations are with one or two orders of magnitude higher than the other HM (with mean value 650 $\mu g/g$). Maximum Fe values are obtained for *C. vagabunda, C. coleothrix* and *C. gracilis,* while in the other three species Fe content is three times lower (Figure 2). A higher value of Fe is observed in *E. intestinalis* from Sinemoretz and Rezovo (autumn 1996 and summer 2000, respectively). The lowest Fe concentration was measured in *U. rigida* species from Ravda (spring 1998) and at Ahtopol (autumn 1996).

If mean Fe values are compared in all green algae depending on the location, a tendency is obtained of increasing Fe content from north to south. If Fe values are plotted for each green species vs. location, it is clear that the observed north-south tendency is mainly due to *Enteromorpha* species whose values increase southwards. Fe concentration for the other green species is more constant with geographical location (e.g., mean Fe values vs. location for *U. rigida* are in the range 300 ± 170 $\mu g/g$ for the whole Black Sea coast).

Sustainable Environment – Monitoring of Radionuclide and Heavy Metal Accumulation in Sediments, Algae
and Biota in Black Sea Marine Ecosystems

69

The results for Mn vary in more narrow interval than Fe (mean value 84 $\mu g/g$). Low Mn content is measured for *U. rigida, E. intestinalis* and *B. plumosa,* while the highest is obtained for *C. coleothrix.* Mn concentrations in different regions change in the following order:

Ahtopol < Kaliakra < Shabla < Tuzlata ≈ Ravda < Sinemoretz < Rezovo.

The mean Mn values for Tuzlata, Ravda and Sinemoretz are close and it is evident that there is no geographic dependence for Mn content. Our data show that green algae from Kaliakra (north) and Ahtopol (south) sites accumulate Fe and Mn to the lowest extent while the highest content is measured at Rezovo (south).

The high Fe and Mn biosorption, compared to the other HM, is connected with their function and major role in the metabolytic processes in marine organisms.

The trace element Cu (like Fe) belongs to the group of biologically important metal ions. Trace metals should be monitored because they play an important role in metabolism and their high or low concentrations can be equally harmful to the living organisms. Cu, Pb and Cd content in green algae are presented in Figure 3.

The Cu data interval is wider compared to Pb and Cd but if mean values ($\mu g/g$) for all algae are compared, we get for Cu 5.6 ± 0.5, Pb 3.3 0.3 and Cd 1.1 ± 0.2. The accumulation patterns sequence is the same for all green algae except *C. coleothrix* where Pb prevails.

Cu mean values are relatively constant along the whole Bulgarian coast (unlike Fe and Mn), and the low Cu content in the environment means that there is no contamination in the marine ecosystems with Cu. The same is true for Pb and Cd whose mean value variations are also small. These results can be explained with the lack of industrial pollution along the coast, except close to the big cities (ports) of Burgas and Varna. The studied locations in this paper are outside the dwellingplaces and this is done in order to obtain the characteristic background values for the measured HM concentrations along the whole coast.

The highest Cu content is measured in *E. intestinalis* from Rossenetz –148$\mu g/g$, which is due to the known anthropogenic contamination of the copper mine in the vicinity. The synergism between Cu and Fe is clearly demonstrated in Rossenetz as Fe value is also high (4890 $\mu g/g$) while the Pb and Cd values are normal.

The behavior of Pb in water ecosystems is complex and its concentration in a great number of natural waters is not higher than 1 $\mu g/g$. Pb is found in seawaters mainly in the form of different organic compounds. Pb content in the studied Black Sea alga species varies in a more narrow interval than Cu.*C. gracilis, C. vagabunda, E. intestinalis* and *B. plumosa* species accumulate Pb in a rather similar way. Pb content variations along the coast are small (like Cu) which also means lack of contamination with Pb.

The determination of Cd content is an important task for the monitoring of HM in marine ecosystems. Cd is poisonous for living organisms even in low concentrations, so it is a hazardous anthropogenic contaminant that should be controlled. It can be concluded from the data in Figure 3 that Cd is present in green algae in comparatively low concentrations – from 0.2 to 3.2 $\mu g/g$ dry weight. The lowest Cd content is in the southern region Sinemoretz, but as a whole the concentration range is narrow in all sites with no geographic dependence. Judging from the alga type, the highest degree of Cd accumulation is found in *C. coleothrix,* while the lowest in *Bryopsis* and *U. rigida.*

Data were measured for Zn and Cr content in some of the studied green algae and the obtained mean values for Zn in *Ulva, E. intestinalis* and *B. plumosa* is 15 $\mu g/g$ (*C. vagabunda* – 23 $\mu g/g$) while for Cr in *Ulva, C. vagabunda* and *C. gracilis* – 1.3 $\mu g/g$ is obtained (*E. intestinalis* – 3.1 $\mu g/g$). The Zn and Cr results for Black Sea macroalgae confirm the lack of HM pollution (like Cu, Pb and Cd) along the Bulgarian coast.

The correlation between accumulation levels of HM concentrations is an important factor for evaluation HM behavior in biota and the determining of these correlations. The coefficient data for *Enteromorpha* and *Ulva* macroalgae show negative correlations between Cd and all measured metals in the two algae species. Pb also correlates negatively with all metal ions in *Enteromorpha* and only with Cu in *Ulva*.

Significant positive correlation coefficient (synergistic interaction between HM) was obtained only for the pair Fe–Mn in *Enteromorpha* while negative correlation (antagonistic) was obtained for Cu–Cd in *Ulva*. These correlation coefficients differ from unit and therefore the correlation dependence is not clearly expressed, meaning that the variables are connected but with weak functional dependence.

All Black Sea green algae data are subjected to cluster analysis for all toxic metal accumulation. The obtained tree diagram for all HM (Figure 9) shows that the algae are combined in two main groups plus *C. coleothrix*. The first main group consist of *Ulva* and *Bryopsis* linked by *Chaetomorpha*. The second group is *Enteromorpha* and *C. vagabunda*.

If the cluster analysis is performed for the toxic elements Cu, Pb and Cd, again two main groups are obtained – first group containing *C. coleothrix* and *Chaetomorpha*. The second group is divided in two – *Enteromorpha, Bryopsis, C. vagabunda* linked with *Ulva*. The main influence on these clusters is due to the presence of Cu because if we exclude Cu from the cluster process the obtained tree diagrams are less affected.

If we assess the obtained algae data depending on the chemical nature of the HM, the obtained tree diagram clearly separates Fe and Cu while Pb and Cd are strongly linked which is clear as these two elements belong to one and the same group of the Periodic table.

Data were obtained for Fe, Mn, Cu, Pb and Cd content in the most widespread Black Sea green macroalgae for the period 1996–2003 (Figs. 9-13) HM environmental behavior in the marine environment of eight locations (Shabla, Tuzlata, Kaliakra, Ravda, Rossenetz, Ahtopol, Sinemoretz, Rezovo), distributed along the whole Bulgarian coast, has been studied. All obtained results prove the dependence of toxic metal accumulation on green algae species as well as on the location.

It can be concluded that Fe, Mn, Cu, Pb and Cd concentration in Black Sea green macroalgae decrease during the studied period. *U. rigida* species accumulate the lowest concentrations of the studied metals. The highest Fe, Mn, Pb and Cd content has been measured in macrophytes from Tuzlata and Rezovo. High Cu concentration is observed in the southern coastal area – Ahtopol and Sinemoretz (the highest at Rossenetz).

All Black Sea green algae data are subjected to cluster analysis for all toxic metal accumulation. The obtained tree diagram for all HM (Figure 15) shows that the algae are combined in two main groups plus *C. coleothrix*. The first main group consist of *Ulva* and *Bryopsis* linked by *Chaetomorpha*. The second group is *Enteromorpha* and *C. vagabunda*.

Sustainable Environment – Monitoring of Radionuclide and Heavy Metal Accumulation in Sediments, Algae and Biota in Black Sea Marine Ecosystems

71

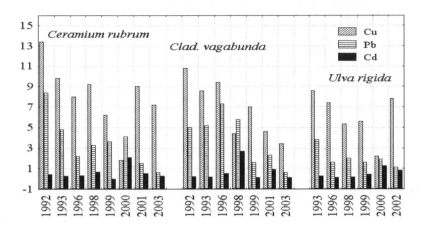

Fig. 13. Seasonal variations of HM concentrations (mg/g) in three alga species

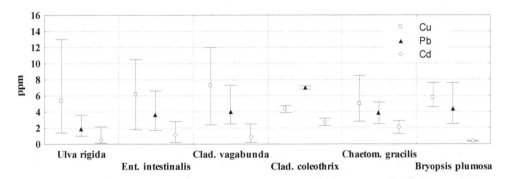

Fig. 14. Mean Cu, Pb and Cd content (µg/g) in different algae species

If the cluster analysis is performed for the toxic elements Cu, Pb and Cd, again two main groups are obtained – first group containing *C. coleothrix* and *Chaetomorpha*. The second group is divided in two – *Enteromorpha, Bryopsis, C. vagabunda* linked with *Ulva*. The main influence on these clusters is due to the presence of Cu because if we exclude Cu from the cluster process the obtained tree diagrams are less affected.

If the obtained algae data are assessed depending on the chemical nature of the HM, the obtained tree diagram clearly separates Fe and Cu while Pb and Cd are strongly linked which is understandable as these two elements belong to one and the same group of the Periodic table.

In conclusion it can be pointed out that Chlorophyta and Rhodophyta algae phyla can be used as bioindicators for monitoring of the ecological state of the Black Sea environment. A comparative analysis of contaminants in different Bulgarian coastline regions leads to the conclusions that the data obtained for red and green macroalgae illustrate the level of contamination by HM and nuclides at seven locations of the Black Sea Coast.

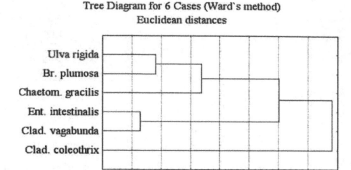

Fig. 15. Tree diagram for heavy metal content in green algae species

The studies of Black Sea macroalgae for marine environmental showed that there is no strict seasonal or local dependence of hazardous element content. All results seem to depend on the biological specificity of the algae and all data show a lack of serious pollution in areas without direct human impact along the Bulgarian Black Sea coast.

All obtained results show that use of macroalgae in marine environmental monitoring reduces the need for complex studies on chemical speciation of aquatic contaminants.

8. Heavy metal levels in Black Sea organisms

Topcuoglu et al. (2003a) investigated the metal content in macroalgae samples collected from the Black sea Turkish coast in the period 1998 to 2000. According to the findings of this study, the heavy metal pollution decreased in Turkish coast of the Black sea in the investigated years (Table 2). In another study it was determined that the Turkish Black sea coast was subjected to heavy metal pollution and the metal concentrations in macroalgae, sea snails and mussels were very high. (Topcuoglu et al., 2002). Romeo et al., (2005) studied the accumulation of trace metal concentrations by measuring them in the mussel collected in the Black sea. The authors found that Cd, Cu, Zn, Hg, Fe, Mn concentration changed between 0.96-1.74 µg/g, 6.64-8.05 µg/g, 108-190 µg/g, 26-33 ng/g, 95-106 µg/g and 14.5-24.5 µg/g mussels, respectively.

Tuzen et al., (2009) measured the trace element concentration in different marine algae (*Antithamnion cruciatum* and *Phyllophora nervosa*) collected from middle and east Black Sea. They found that the highest trace element concentration was determined for iron and the lowest for cadmium (Table 2). The authors suggested that the marine algae samples should be analyzed more often in Turkey with respect to toxic elements. Edible marine algae samples could be used as a food supplement to help meet the recommended daily intakes of some mineral and trace elements.

Zinc, copper, cadmium, lead and cobalt concentrations in Mediterranean mussel *Mytilus galloprovincialis* and sea snail *Rapana venosa* from the Sinop coasts of the Black sea have been measured (Turk-Culha et al., 2007). Significant differences were found in metal concentrations between the species. Similar significant differences observed with regard to different metals. The concentrations of Pb, Cd and Co were determined under detection limit for the species. The other metal levels in the Mediterranean mussel and the sea snail were significantly higher than those in fishes.

From 1996 to 2002, Fe, Mn, Cu, Pb and Cd distribution in six green macroalgae species from the Bulgarian Black Sea coast were determined by Strezov et al, (2005a). For all algae, average heavy metal concentrations were 650 µg/g for Fe, 184 µg/g for Mn, 5.6 µg/g for Cu, 3.3 µg/g for Pb and 1.1 µg/g for Cd. These data show that heavy metal contents in different species demonstrate various degree of metal accumulation. The obtained higher values in the northern part of the studied zone can be attributed to the discharge influence of the big rivers entering the Black sea, such as Danube, Dnyeper, Dnyester, and local pollutant emissions, as well. The obtained data also show that there is no strong contamination in green macroalgae with heavy and toxic metals along the whole Bulgarian Black sea coast.

Metal contamination in the Black sea alga species (green and red) was studied from 1992 to 2003, using radionuclide approach (Strezov et al, 2005b). They found that radionuclide and metal concentrations depend on the macrophyte nature and all data show the lack of strong pollution along the Bulgarian Black sea coast.

9. Data comparison between neighboring seas

The comparison of the data (Tables 6 – 8) for radioecological status of the Black sea marine ecosystems and the Mediterranean marine ecosystems concerning the radionuclide and trace metal content indicated no strong anthropogenic pollution along the Bulgarian shore as the Mediterranean data are higher. When comparing the accumulation of HM in one and the same algae the corresponding data for Black sea algae are lower than those from the Marmare Sea or the Mediterranean.

Species	Metals						Reference
	Cd	Co	Zn	Mn	Pb	Cu	
Mussels Sea snail (soft	<0.02- 2.01-	<0.05-5.36 <0.05-0.7	78.12- 73.3-	5.66-22.8 3.90-	<0.05-2.60 <0.5	7.21 36.19	Topçuoğlu et al. (2002)
*Ulva lactuca 1998/1999	<0.02/ <0.02	<0.05/ <0.05	21.2/ 9.6	45.1/ 21.8	<0.1/ <0.1	13.8 3.87	
* Cystoseira 1998/1999	<0.02/ <0.02	<0.05/ <0.05	35.1/ 13.9	32.1/ 6.7	<0.1/ <0.1	5.7 2.2	
* Pterocladiella capillacea	1.53/ 1.36	<0.05/ <0.05	119.8/ 86.2	91.1/ 52.1	<0.1/ <0.1	10.3 5.3	Topçuoğlu et al. (2003)
**Ulva lactuca 1998/1999	<0.2/ <0.2	<0.05/ <0.05	13.5/ 394.4	41.1/ 12.5	<0.1/ <0.1	11.3 7.7	
** Cystoseira 1998/1999	<002/ <0.02	<0.05/ <0.05	43.9/ 191.5	27.3- 22.7	<0.1/ <0.1	593 590	
**Pterocladiela capillacea 1999	<0.02	<0.05	176.8	10.8	<0.1	<0.03	
Cystoseira c. Cystoseira b.	0.22-1.87 0.10-1.25	- -	- -	9-60 9-54	0.4-6.3 0.7-2.4	0.3-6.0 0.6-9.0	Strezov-Nonova (2003)
Mussels	<0.02	-	312-396.5	46.9-73.05	<0.05-108	11.75-23.25	Bakan & Böke-Özkoç (2008)

*West Black Sea: ** East Black Sea: ^ µg/kg

Table 6. Heavy metal levels (µg/g dry wt) in some living organism from the Black Sea

Algae	Cd	Cr	Cu	Mn	Zn	Pb	Reference
Ulva lactuca	1.35	< 1	7.5	-	34.2	6.5	Guven, 1998
Cystoseira barbata	1.3	< 1	4.2	-	33	5.3	Guven, 1998
Ulva lactuca	0.5	0.5	24	50	24.1	23.5	Topcuoglu, 2001a
Cystoseira barbata	0.75	0.95	6.85	25	97	14	Topcuoglu,2001b
Ceramium rubrum	0.8	1.5	16	59	62	11	Topcuoglu 2001
Ulva sp	0.24	4.78	4.7	194	26.1	-	Malea, 2000
Ulva sp	0.6	1.56	5.5	-	5.2	3.68	Muse, 1999
Enteromorpha sp.	0.07	0.54	11.4	21	14	1.06	Favero, 2002
Ulva sp	0.18	1.63	5.8	-	45	1.94	Conti, 2003
Padina pavonica	1.56	3,6	13.3	-	84	11.4	Campanella 2001
Ulva sp	0.8	1.8	5.6	40	24	1.7	Strezov, 2005a
Ceramium s	0.9	6	7.6	120	22	2.2	Strezov, 2005a
Cladophora sp.	1	7	6	170	19	3.5	Strezov, 2005a
Enteromorpha sp.	0.8	5.3	7	47	14	2.4	Strezov, 2005a
Cystoseira sp.	0.3	2.3	4	42	1.6	18	Strezov, 2005b
Chaetomorpha sp	1.3	7	5	180	12	2.7	Strezov, 2005b
Corallina sp.	0.7	4.8	15	55	13	1.4	Strezov, 2005b
Callithamnion sp	0.5	3.7	5.4	87	18	2.3	Strezov, 2005b

Table 7. Trace metal content (mg/kg) in marine algae from the Black and Mediterranean seas

Algae	137Cs	40K	210Pb	226Ra	Reference
Ceramium rubrum	0.8	-	-	-	Bologa (1996)
Cystoseira sp.	15	900	-	-	Guven (1993)
Cladophora sp.	n.d.	2170	-	-	Guven (1993)
Enteromorpha sp.	5	1076	-	-	Guven (1993)
Chaetomorpha sp.	11	2525	-	-	Guven (1993)
Ulva sp.	6	930	-	-	Guven (1993)
Corallina sp.	5	250	-	-	Guven (1993)
Ceramium rubrum	12	880	-	-	Guven (1993)
Ulva lactuca	< 3	-	-	-	Topcuoglu (2001a)
Cystoseira barbata	5.9	-	-	-	Topcuoglu (2001b)
Ulva sp	< 1.2	900	3.49	< 1.7	Al-Masri (2003)
Cystoseira sp	< 1.1	1800	8	1.2	Al-Masri (2003)
Ulva lactuca	7.2	-	-	-	Othman (1994)
Ulva sp.	3.4	596	6	9.3	Strezov, (2005a)
Ceramium sp	9.4	1343	13	17	Strezov, (2005b)
Cladophora sp.	12	1300	8	12	Strezov, (2005a)
Enteromorpha sp.	4.5	690	7	10	Strezov, (2005a)
Cystoseira sp.	5.4	1400	12	11	Strezov, (2005b)
Chaetomorpha sp.	2.3	1860	10	7	Strezov, (2005a)
Corallina sp.	2.1	140	12	10	Strezov, (2005b)
Callithamnion sp	4.4	1580	10	7	Strezov, (2005b)

Table 8. Radionuclides mean content in Black sea algae (Bq/kg)

Aquatic organisms, especially macroalgae, are widely used as bioindicators for the study of marine contamination by radionuclides and heavy metals. Some species tolerate high levels of pollutants and can successfully be used to obtain reliable information of marine ecological status. In order to evaluate the ecological status of coastal habitat and provide valuable data for estimation of contamination with radionuclides in Bulgarian Black sea, a monitoring program was started and performed along the coast. Eleven macroalgae species were monitored by collecting samples from 20 reference locations during the 1996-2005 period.

In this way the accumulation of hazardous contaminants was traced over a long period of time and tendencies for the behaviour of those elements were evaluated. The same comparison was made for radionuclides (Table 8) in the studied algae species and Black Sea Cs content was found to be similar to the other basins while natural nuclides were slightly higher in the Black sea.

10. Conclusion

Radionuclide and HM accumulation capacity has been studied in Black sea sediments three algae phylum along the Bulgarian Black Sea coast during the period 1996 – 2004. The natural isotope concentrations are higher than technogenic ones and red alga *Ceramium rubrum* shows the highest level of nuclide concentrations. The status of the marine environment in all studied areas was evaluated by cluster analysis of macroalgae data from all geographic zones. Analysis results show logical geographic dependence of contaminant content and locations of higher content are distinctly separated from those of clean areas.

The full scale monitoring done on the whole Bulgarian Black Sea coast resulted in collecting information for the different equilibrium processes, taking place in the coastal regions which govern the radioactive pollution of rivers, adjacent salt lakes, sea sediments and water and other harmful effects of human activities but also their behaviour in the marine ecosystems (algae, sea mussels, fish and other marine organisms the rates of exchange and the pathways towards man). It can be pointed out that Chlorophyta and Rhodophyta algae phyla can be used as bioindicators for monitoring of eco-toxicological state of the Black Sea environment. A comparative analysis of contaminants in different Bulgarian coastline regions leads to following conclusions:

- A data base for isotope accumulation, sorption and migration of nuclides and HM is created to help the future assessment and biosphere in whole Bulgarian coastal zone.
- The data obtained for red and green macroalgae illustrate the level of contamination at the locations of the Black Sea Coast
- There is no strict seasonal or local dependence of hazardous element content. All results seem to depend on biological specificity of the algae
- All data show the lack of serious pollution along the Bulgarian Black Sea coast.
- All obtained results for sediments and macroalgae in marine environmental monitoring reduces the need of complex studies on chemical speciation of aquatic contaminants and makes algae valuable indicators for the seawater quality assessment.
- Modelling the transfer processes of radionuclides in environment and the different pathways of isotope migration in the marine environment to predict the potential hazard for the population

11. References

Al-Masri M, S. Mamish, Y. Budier. 2003. Radionuclides and trace metals in eastern Mediterranean Sea algae, *J. Env. Radioact.* 67, 157-168

Babich, H., Devaris, M., Stotzki, G.: 1985, 'The mediation of mutagenicity and clastogenicity of heavy metals by physiochemical factors', *Environ. Res.* 37, 325–331.

Bakan, G. and Böke-Özkoç, H. 2008. An ecological risk assessment of the impact of heavy metals in surface sediments on biota from mid-Black sea coast of Turkey. Int. *Journal of Environmental Studies*, 64: 45-57.

Baumann, M., Segl, M., Wefer, G., von Bodungen, B., 1989. Chernobyl Derived Radiocaesium in the Norwegian Sea: Flux between Water Column and Sediment. In The Radioecology of Natural and Artificial Radionuclides *XV Regional Congress of IRPA*, Gotland Sweden, 318-323.

Bologa A., M. Apa, A. Cociasu, E. Cuingoglu, V. Patrascu, V. Piescu, L. Popa, 1996. Present level of contaminants in the Romanian Black Sea sector, *IAEA TechDoc 1094*, 469-475

Broberg, A., 1989. Distribution of Cs-137 in Lake Sediments. In *Radioecology of Natural & Artificial Radionuclides XV Regional IRPA Congress* , Gotland, Sweden, 301-305

Buesseler, K.O., Livingston, H.D., Hohjo, S., Hay, B.J., Manganini, S.J., Degens, E., Ittekot, V., Izdar, E., Konuk, T., 1987. Chernobyl Radionuclides in a Black Sea Sediment Trap. *Nature* 329, 825-828.

Coban, B., N Balkis,. A. Aksu. 2009. Heavy metal levels in sea water and sediment of Zonguldak, Turkey. *Journal of Black Sea/Mediterranean Environment*, 15, 23-32.

Conti M. E., G. Cecchetti. 2003. A biomonitoring study: trace metals in algae and molluscs from Tyrrhenian coastal areas, *Environ. Res.* 93, 99-112

Ergul, H.A., S. Topcuoglu, E. Olmez, C. Kirbasoglu. 2008. Heavy metals in sinking particles and bottom sediments from the eastern Turkish coast of the Black Sea. Estuarine, *Coastal and Shelf Science*, 78, 396-402

Favero N., M. Frigo. 2002. Biomonitoring of metal availability in the southern basin of the lagoon of Venica (Italy) by means of macroalgae, *Wat. Air Soil Poll.* 140, 231-246

Förstner, U., W. Salamons. 1980. Trace metal analysis on polluted sediments, I: assessment of sources and intensities. *Environmental Technology Letters 1, 494-505.*

Güven, K.C., Topcuoglu, S., Kut, D., Esen, N., Erentürk, N., Saygi, N., Cevher, E., Güvener, B., Öztürk, B. (1992). Metal uptake by Black Sea Algae. *Bot. Marina*, 35, 337-340.

Guven K., Topcuoglu S., Gungor. N. 1993. Chernobyl radioactivity in Algae Collected from the Marmara and Black Sea. *Turkish J. Nucl. Sciences* 20, 21-30

Guven K., Okus E., Topcuoglu S., Esen N., Kucukcezzar R., Seddingh E., Kut. D. 1998. Heavy Metal Accumulation in Algae and Sediments of Black Sea coast of Turkey *Tox. Env. Chem.*, 67, 435-440

Hadderingh, R. H., Kema, N.V. (1989). Distribution of Cs-137 in the Aquatic Food-chain of Ijsselmeer after the Chernobyl Accident. In *The Radioecology of Natural and Artificial Radionuclides. XV Regional Congress of IRPA*, Gotland Sweden, 325-330

Ignatov, E. 2008. Coastal and Bottom Topography, the Black Sea Environment, Springer Verlag Berlin Heidelberg, 62 pp.

Kalugina – Gutnik A.A. (1975). Phytobentos of the Black Sea. *Naukova dumka Kiev, p. 16.*

Kilgore, J., Dibble, E., Hoover, J. 1993. Relationships between fish and aquatic plants: A plan of study, *Miscellaneous paper A-93-1*, U.S. *Army Corps of Engineers WES*, Vicksburg, MS.

Kulebakina, L.G., Zesenko, A.J. (1984). Dynamics of contamination levels of ^{90}Sr and ^{137}Cs in water and hydrobionts in the Danube River delta and surrounding part of the Black Sea. Radiochemical Ecology & Problems of Contamination Kiev, *Naukova Dumka*, 16-40.

Kulebakina, L., Egorov, V., Polikarpov, G. 1988. Redistribution and of ^{137}Cs in Black Sea water in 1986 - 1988 (in Russian), *Gidrobiologicheski Journal* 24, 69.

Kulebakina, L. and Polikarpov, G., 1989. Radiological Investigation of Black Sea Ecosystems after the Chernobyl Accident. *Proc. Conference on Radioecology High Tatras, Chechoslovakia, 131-155.*

Keondjan V.P, Kudin A.M., Terehin Yu., Kiev, 1990. Practical Ecology of Sea Regions. The Black Sea, *Naukova Dumka, 46 - 56*

Kiratli N., M. Ergin. 1996. Partitioning of heavy metals in surface Black Sea sediments, *Applied Geochemistry*, 11, 775-788.

Malea P, S.Haritonidis. 2000. Use of green alga *Ulva rigida* as an indicator species to reassess metal pollution in Thermaikos Gulf, Greece, after 13 years *J. Appl. Phicology*, 12, 169-176

Manahan, S. 1999. Environmental Chemistry, Seventh edition, Lewis Publishers, New York, London, Washington, pp 783.

Muse J., J.Stripeikis, F.Fernandez, L.Huicqui, M.Tudino, C.Carducci, O.Troccoli. 1999. – Seaweeds in the assessment of heavy metal pollution in the Gulf San Jorge, Argentina *Env. Poll.*, 104, 315-322.

Nijenhuis I., H. Bosch, J. Sinninghe, H. Burmsack, G.De Lange. 1999. Organic matter and trace element rich sapropels and black shales: a geochemical comparson. *Earth and Planetary Science Letters*, 169, 277-290.

Othman I., J.Yassine, I.Bhat. 1994. Measurements of some radionuclides in the marine coastal environments of Syria. *Sci Tot. Env.* 53, 57-60

Polikarpov, G. 1966. Radioecology of Aquatic Organisms, *N. Holland Publ. Company*, Amsterdam.

Polikarpov G.G., Kulebakina L.G., Timoshchuk V.I. and Stokozov N.A. (1991). ^{90}Sr and ^{137}Cs in surface waters of the Dniepr River, the Black Sea and the Aegean Sea in 1987 and 1988. *J. Environ. Radioactivity*, 13, 25-34.

Rainbow, P. 1995. Biomonitoring of heavy metal availability in the marine environment. *Mar. Poll. Bull.* 31, 183-192

Rainbow, P. S. Phillips, D. 1993. Cosmopolitian biomonitors of trace metals, *Mar. Pollut. Bull.* 26, 593–601.

Roeva, N. N., Rovinski, F. and Kononov, E.: 1996, 'Specific features of heavy metal behavior in different nature media', *Zhurnal Analytichnoy Khimii* 51, 384–391.

Romeo M., C. Frasila, M. Gnassia-Barelli, G. Damiens, D. Micu, G. Mustata. 2005. Biomonitoring of trace metals in the Black Sea (Romania) using mussels *Mytilus galloprovincialis. Water Research*, 39, 596-604.

Sokolova, I. 1971. Calcium, ^{90}Sr and Strontium in Marine Organisms (in Russian), Kiev, *Naukova Dumka, 239-246*

Strezov, A., M. Milanov, P. Mishev, T. Stoilova. 1998. Radionuclide accumulation in near-shore sediments along Bulgarian Black Sea coast. *Appl. Radiat Isotopes 49 (12), 1721-1728*

Strezov, A., T.Stoilova, A. Jordanova, N. Petkov. 1999. Determination of Cs and Natural Radionuclide Concentrations in Sediments, Algae & Water. *Water Sci. Tech.* 39, 21-26.

Strezov A., T. Nonova. 2003. Monitoring of Fe, Mn, Cu, Pb and Cd levels in two brown macroalge from the Bulgarian Black sea coast. *Int. J. Env. Analyt Chemistry*, 83, 1045-1054

Strezov A., T. Nonova. 2005a. Environmental Monitoring of Heavy Metals in Bulgarian Black Sea Green Algae. *Environmental Monitoring and Assessment*, 105, 99-110.

Strezov A, T. Nonova. 2005b. Comparative analysis of heavy metal and radionuclide contaminants in Black Sea green and red macroalgae, *Water Science & Technology*, 51 1-8.

Shvedov ,V., Ivanova, L., Maximova, A., Stepanov, A. (1962). Radioactive Substances Content and Distribution in Sea Water. In *Radioactive Contamination of the Environment*, ed. Shvedov V. and Shirokov S., Moscow, 236-242

Topcuoglu, S., Esen, N., Egilli, E., Gungor, N, Kut, D., 1998. Trace Elements and 137Cs in Macroalgae and Mussels from the Kilyos in Black Sea, *Proc. Int. Symposium on Marine Pollution Monaco 1998*, IAEA SM-354 pp. 283-284.

Topcuoglu, S., N. Gungor. 1999. Radionuclide Concentrations in Macroalgae and Sediment Samples from the Bosphorus, *Turk. J. Mar. Sci.* 5, 19-29.

Topcuoglu S, D. Kut, N. Esen, N.Gungor, E.Olmez, C Kirbasoglu. 2001. 137Cs in biota and samples from Turkish coast of the Black Sea, 1997-1998. *J Rad. Nucl Chem* 250, 381-384

Topcuoglu S., K. Guven, C.Kirbasoglu, N.Gungor, S.Unlu, Z.Yilmaz. 2001. Heavy Metals in marine algae from Sile in the Black Sea *Bull. Env. Contam. Toxicol.*, 67, 288-294

Topcuoglu S., C.Kirbasoglulu, N.Gungor. 2002. Heavy metals in organisms and sediments from Turkish coast of the Black Sea, 1997-1998. *Environmental International*, 27, 521-526.

Topcuoglu S., K. Guven, N.Balkis, C.Kirbasoglu, 2003a. Heavy metal monitoring of marine algae from the Turkish coast of the Black Sea, 1998-2000. *Chemosphere*, 52, 1683-1688.

Topcuoglu S., H.Ergul, A.Baysal, E.Olmez, D.Kut. 2003b. Determination of radionuclide and heavy metal concentrations in biota and sediment samples from Pazar & Rize stations in eastern Black Sea. Fresenius Environmental Bulletin,12, 695-699.

Topcuoglu S., E. Olmez, C. Kirbasoglu, Y. Yilmaz, N. Saygin. 2004. Heavy metal and radioactivity in biota and sediment samples collected from Unye in the eastern Black Sea. *Proceedings 37th CIESM Congress*. Barcelona, Spain, 250 pp.

Tuncer G, T.Karakas, T.Balkas, C.Gokcay, S.Aygnn, C.Yeteri, G.Tuncel. 1998. Land based sources of pollution along the Black Sea Coast of Turkey: Concentrations and Annual Loads to the Black Sea. *Marine Pollution Bulletin*, 36, 409-423.

Turk-Culha S., L. Bat, M. Culha, A. Efendioglu, M. Andac, B. Bati. 2007. Heavy metals levels in some fishes and molluscs from Sinop., Peninsula of the Southern Black Sea, Turkey, . Rapport , *38th CIESM Congress Proceedings* , 323 pp.

Tuzen, M. 2009. Toxic and essential trace elemental content in fish species from the Black Sea, Turkey. *Food and Chemical Toxicology*, 47, 1785-1790.

Zaitzev Y. 1992. Ecological state of the Black Sea shelf zone, Ukrainian coast. Review, *Hydrobiol, Journal*, 28 (4), 36.

Part 2

Occupational Exposure of Environmental Contaminants

5

Environmental Contaminations and Occupational Exposures Involved in Preparation of Chemotherapeutic Drugs

Shinichiro Maeda[1,2], Masako Oishi[1],
Yoshihiro Miwa[1] and Nobuo Kurokawa[1]
[1]*Department of Pharmacy, Osaka University Hospital*
[2]*Graduate School of Pharmaceutical Sciences, Osaka University*
Japan

1. Indroduction

Many healthcare workers are concerned about the risk of occupational exposures to hazardous drugs. Since Falck [Falck et al.,1979] reported that mutagens were detected in the urine samples of nurses involved in chemotherapy, many reports have been published about the presence of urinary mutagen [Roth et al.,1994; Burgaz et al.,1999; Kasuba et al.,1999; Lanza et al.,1999; Jakab et al.,2001; Kopjar & Garaj-Vrhovac,2001], and about the detection of unchanged hazardous drugs in the urine samples of healthcare workers [Hirst et al.,1984; Sessink et al.,1992, 1994, 1995; Ensslin et al.,1994; Minoia et al.,1998; Burgaz et al.,1999; Pethran et al.,2003]. Tomioka [Tomioka & Kumagai,2005] monitored occupational exposures to hazardous drugs by routes of exposures and the levels at which the drugs exert their effects; i.e. (i) external exposures, exposure to airborne drugs and drugs deposited on the working table; (ii) internal exposures, presence of drugs or their metabolites in blood and urine; (iii) cellular level effects, presence of mutagens in urine and frequency of sister chromatid exchanges; and (iv) effects on individual level, susceptibility to cancer and effects on reproduction. Recent improvements of analytical instruments have permitted direct monitoring of external exposures and internal exposures.

Hospital pharmacists in Japan have been required to prepare chemotherapeutic drugs. The Japanese Society of Hospital Pharmacists (JSHP) revised the "Guideline for the Handling of Antineoplastic Drugs in Hospitals" in 2005, and recommended standard precautionary measures for using the laminar flow cabinet, masks, gloves, caps, disposable nonwoven clothes, and luer-lok syringes. This guideline was based on the "NIOSH alert" by the National Institute for Occupational Safety and Health [NIOSH,2004]. In addition, the JSHP investigated occupational exposures to cyclophosphamide in several hospitals, and reported that internal exposures and external exposures varied greatly in individuals, in hospitals and even in countries. This investigation led to increased concerns among hospital pharmacists about the exposures to chemotherapeutic drugs.

Before the JSHP investigation, we have investigated whether the precautionary measures we took were adequate to prevent occupational exposures [Ikeda et al.,2007; Maeda et al.,2010]. We focused on occupational exposures to epirubicin, cyclophosphamide and ifosfamide, and reported that the surfaces of biological safety cabinets (BSCs) and ambient environments were contaminated by these drugs during preparation by pharmacists (external exposures). However, detection frequencies and amounts of these drugs were low levels, compared with previous reports in Japan [Nabeshima et al.,2008; Yoshida et al.,2009]. In addition, no drugs were detected in sera and urine samples of healthcare workers involved in chemotherapy (internal exposures). Thus, we concluded that adequate precautionary measures and improved awareness regarding handling of chemotherapeutic drugs could reduce the risk of occupational exposures. These investigations were focused on specific drugs, however, a progress of chemotherapy increases amounts of and varieties of chemotherapeutic drug uses. Thus, more versatile methods are desired.

Multicomponent analyses are useful for monitoring environmental contaminations, such as pesticide chemicals [Lissalde et al.,2011; Miao et al.,2011] and air pollution [Skoczynska et al.,2008; Sebok et al.,2009]. However, there are only several reports in the fields of occupational exposures to healthcare workers [Larson et al.,2003; Sabatini et al.,2005; Sottani et al.,2007, 2008; Nussbaumer et al.,2010].

In this paper, we used liquid chromatography-mass spectrometry/mass spectrometry (LC-MS/MS) and developed multicomponent analysis procedures for chemotherapeutic drugs, and assessed the levels of environmental contaminations involved in preparation of chemotherapeutic drugs. We selected ten kinds of drugs mainly used in chemotherapy; cyclophosphamide, ifosfamide, epirubicin, doxorubicin, vindesine, vincristine, vinblastine, irinotecan, docetaxel and paclitaxel. Classification of drugs, chemical natures, regulations, uses of drugs, side effects, carcinogenicities, cytotoxicities [Goolsby & Lombardo,2006] and pregnancy categories of each drug were summarized in Table 1.

No. Name of drugs	Regulations	Classification of drugs (chemical natures)	Main uses of drugs
1 Vindesine	Powerful drug	Vincalkaloid (antimicrotuble drug)	Leulemia, Lung cancer
2 Vincristine	Powerful drug	Vincalkaloid (antimicrotuble drug)	Leukemia, Malignant lymphoma, Multiple myeloma
3 Vinblastine	Powerful drug	Vincalkaloid (antimicrotuble drug)	Malignant lymphoma, Urothelial carcinome
4 Doxorubicin	Powerful drug	Anthracycline (anticancerous antibiotic)	Malignant lymphoma, Breast cancer, Lung cancer
5 Epirubicin	Powerful drug	Anthracycline (anticancerous antibiotic)	Leukemia, Malignant lymphoma, Breast cancer
6 Ifosfamide	Powerful drug	Mustard (alkylating drug)	Lung cancer, Prostatic cancer
7 Cyclophosphamide	Powerful drug	Mustard (alkylating drug)	Leukemia, Breast cancer, Rheumatism
8 Irinotecan	Powerful drug	Topoisomerase I inhibitor	Lung cancer, Ovarian cancer, Colorectal cancer
9 Docetaxel	Toxicant	Taxane (antimicrotuble drug)	Breast cancer, Lung cancer, Uterine cancer
10 Paclitaxel	Toxicant	Taxane (antimicrotuble drug)	Breast cancer, Lung cancer, Uterine cancer

No.	Name of drugs	Main side effects	Hazard potential		
			Carcinogenicity	Cytotoxicity	Pregnancy
1	Vindesine	Myelosuppression		Vesicant	
2	Vincristine	Peripheral neuropathy, Cardiotoxicity		Vesicant	Category D
3	Vinblastine	Myelosuppression		Vesicant	Category D
4	Doxorubicin	Delayed myelosuppression and cardiotoxicity	Probably carcinogenic	Vesicant	Category D
5	Epirubicin	Delayed myelosuppression and cardiotoxicity		Vesicant	Category D
6	Ifosfamide	Hemorrhagic cystitis		Irritant	Category D
7	Cyclophosphamide	Myelosuppression, Hemorrhagic cystitis	Carcinogenic	Irritant	Category D
8	Irinotecan	Delayed diarrhea, Myelosuppression		Non-vesicant	Category D
9	Docetaxel	Myelosuppression, Pulmonary fibrosis		Vesicant	Category D
10	Paclitaxel	Myelosuppression, Perepheral neuropathy		Vesicant	Category D

Regulations, use of drugs and side effects were authorized by the Ministry of Health, Labour and Welfare in Japan.

Carcinogenicities were classified by IARC monographs on the evaluation of the carcinogenic risk of chemicals to humans.

Pregnancy categories were classified by U.S. Department of Health and Human Services Food and Drug Administration. Category D was "there is positive evidence of human fetal risk based on adverse reaction data from investigational or marketing experience or studies in humans, but potential benefits may warrant use of the drug in pfegnant women despite potential risks."

Table 1. Charactaristics of selected chemotherapeutic drugs.

2. Materials and methods

2.1 Chemicals and materials

Doxorubicin, ifosfamide, cyclophophamide and paclitaxel were purchased from Wako Pure Chemical (Osaka, Japan). 3,4-Anhydro vincristine [internal standard (IS) 1] was purchased from Toronto Research Chemicals (Toronto, Canada). Camptothecin (IS 2) was purchased from Tokyo Chemical Industry (Tokyo, Japan). Docetaxel, carminomycin (IS 3) and trofosfamide (IS 4) were purchased from Santa Cruz Biotechnology (California, USA). Irinotecan and cephalomannine (IS 5) were purchased from Sigma-Aldrich (Missouri, USA). Vincristine, vinblastine and epirubicin were kindly provided by Nippon Kayaku. Vindesine was kindly provided by Shionogi. Acetonitrile and methanol (LC-MS chromasolv) were also purchased from Sigma-Aldrich (Missouri, USA), and formic acid was purchased from Wako Pure Chemicals.

2.2 Preparation of stock solutions, working solutions and calibration standards

Vindesine was prepared in a solution of 0.1% formic acid-water to obtain a final concentration of 2 mg/ml. Camptothecin was prepared in a solution of 0.1% formic acid-methanol to obtain a final concentration of 100 μg/ml. Other drugs were prepared in a solution of 0.1% formic acid-methanol to obtain a final concentration of 2 mg/ml. Aliquots of these solutions were stored at -80°C, then diluted in 0.1% formic acid-methanol to obtain a final concentration of 100 μg/ml each stock solution and stored at -30°C.

Stock solutions of three vincalkaloids were mixed to working solutions, containing 33.3 μg/ml of vindesine, vincristine and vinblastine, respectively. Stock solutions of other seven drugs were also mixed and diluted in 0.1% formic acid-methanol to working solutions, containing 5 μg/ml of doxorubicin, epirubicin, ifosfamide, cyclophosphamide, irinotecan, docetaxel and paclitaxel, respectively. The working solutions of internal standard mixtures were prepared by containing 100 μg/ml of 3,4-anhydro vincrisitine, and containing 10 μg/ml of carminomycin, trofosfamide, camptothecin and cephalomannine, respectively. These working solutions were also stored at -30°C. We used black Eppendorf tubes for stock solutions of anthracycline drugs and vincalkaloid drugs, and for all working solutions. The calibration standards were prepared by diluting these working solutions.

2.3 Chromatographic conditions

An Alliance 2695 HPLC separation module (Waters; Massachusetts, USA) with PDA detector, cooled autosampler and column oven was used to perform this monitoring. Chromatographic separation was achieved on an octadecyl silyl column (Inertsil® ODS-3; 50 mm×2.1 mm; particle size, 3 μm; GL Sciences, Tokyo, Japan) with a guard column (cartridge guard-column E®; 20 mm×2.0 mm; particle size, 3 μm, GL Sciences, Tokyo, Japan). Column oven was maintained at 30°C and autosampler was maintained at 5°C. The mobile phases consisted of 0.1% formic acid-water (mobile phase A) and acetonitrile (mobile phase B). A flow rate was 0.3 ml/min and gradient elution was performed in the following manner: 15% of mobile phase B to 45% over 10 min; 45% of mobile phase B to 80% over 7 min. Subsequently, the concentration of mobile phase B was linearly decreased to 15% for 1 min and equilibrated for 4 min. Total run time was 22 min.

2.4 Mass spectrometry conditions

A tandem quadrupole MS TQD (Waters; Massachusetts, USA), operated in multiple reaction monitoring (MRM) in positive electrospray ionization (ESI) mode, was used for detection and MassLynx 4.1 software was used for data acquisition and processing. MS/MS parameters (precursor ion, product ion, cone energy, collision energy and retention time) of each drug were individually optimized by QuanOptimize software and syringe pump infusion in primary mobile phase by constant flow (Table 2).

2.5 Precautionary measures and personals involved in preparation of chemotherapeutic drugs

Precautionary measures were based on "NIOSH alert" [NIOSH,2004] and "Guidelines for the Handling of Antineoplastic Drugs in Hospitals" by JSHP (Table 3). A preparation area for outpatient chemotherapy was selected as an environmental monitoring. The numbers of person involved in preparation for outpatients were eight; six men and two women, and ages were from twenty-four to forty-one. We rotated schedules regularly, and limited successive preparation time to 1 hour.

Parameters / Compounds	Precersor ion	Product ion	Cone energy (V)	Collision energy (V)	Retention time (min)
Vindesine	754.5	124.2	55	50	1.65
Vincristine	825.4	765.6	70	40	5.02
Vinblastine	811.4	224.1	60	45	5.75
3,4-Anhydro vincristine (IS 1)	807.4	747.5	65	30	6.22
Doxorubicin	544.5	397.2	20	12	6.18
Epirubicin	544.5	397.2	20	12	6.60
Carminomycin (IS 2)	514.2	307.2	20	25	8.29
Ifosfamide	261.0	153.9	45	25	7.21
Cyclophosphamide	261.0	140.3	45	30	7.63
Trofosfamide (IS 3)	323.1	153.9	30	25	12.69
Irinotecan	587.7	167.3	60	45	5.77
Camptothecin (IS 4)	349.1	305.2	50	25	9.51
Docetaxel	808.8	226.1	20	14	15.95
Paclitaxel	854.8	286.3	25	24	16.48
Cephalomannine (IS 5)	832.8	264.2	25	20	16.09

Table 2. Parameters of MS/MS analysis.

No	Precautionary measures
1	Prepare chemotherapeutic drugs in a centralized area restricted to authorized personnel with expertise in the preparatory techniques and the characteristics of these drugs.
2	Prepare these drugs in a biological safety cabinet Class II Type B.
3	Place a superabsorbent sheet on the surfaces of the preparation cabinets.
4	Use syringes with Luer-Lok-type fittings for preparing these drugs.
5	Use gowns made of a lint-free, low-permeability fabric.The gown should have a closed front, long sleeves, and elastic or knit-closed cuffs.
6	Use disposable nitrile rubbers gloves doubly, and change gloves every hour or on accidental exposures.
7	Wear disposable mask and cap.
8	Remove protective clothing carefully to avoid spreading contaminations.
9	Maintain a negative pressure in the drug vials.
10	Always close the cover of trash boxes used to dispose of these drugs.

These measures are based on the NIOSH alert, and we take measure No. 3 for additional precaution.

Table 3. Precautionary measures we took.

2.6 Sampling procedures of ambient environment

We collected wiping samples from the surfaces of the BSCs and from the surfaces of the tables inside a separated area and outside one (Fig. 1). Investigations were carried out as soon as daily preparation for outpatients had finished and before daily cleaning-up procedures, and obtained samples were immediately extracted and measured by LC-MS/MS equipments. We took this monitoring on seven days for successive two weeks.

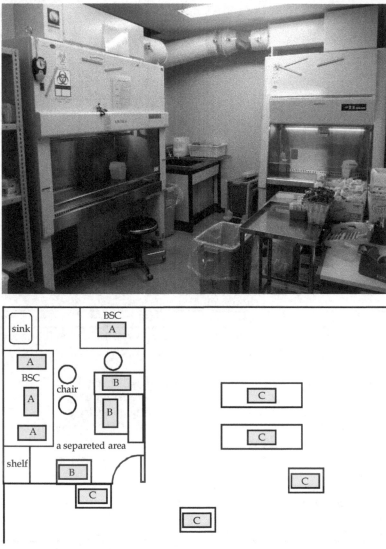

Fig. 1. Photograph of a preparation area (inside a separated area) for outpatients chemotherapy, and floor layout of a preparation area and sampling spots. A) Surfaces of the BSCs, B) Surfaces of the tables inside a separated area, C) Surfaces of the tables outside a separated area.

An extraction of wiping samples for environmental assessments was modified by our previous report [Maeda et al.,2010]. Briefly, we first applied internal standards mixtures, containing 10 μg of 3,4-anhydro vincrisitine, 500 ng of carminomycin, trofosfamide, camptothecin and cephalomannine to sampling spots. After these spots were air-dried, we wiped 800 cm² area (20 cm×40cm) with a sheet of Kimwipe® S-200 (120 mm×215mm, Nippon Paper Crecia, Tokyo, Japan) wetted with 1 ml of 0.1% formic acid-70% methanol. We repeated wiping operation twice, then placed both sheets in light-blocking polypropylene conical tubes and added 8 ml of 0.1% formic acid-70% methanol. We shook these tubes for 30 min, 2,000 rotations per min. Obtained extracts were directly injected into LC-MS/MS equipments without dilution.

3. Results

All drugs were clearly detected and quantified over a total run time of 22 min. The calibration curves were fitted by the linear regression 5-1000 ng/wipe for doxorubicin, epirubicin, ifosfamide, cyclophosphamide, irinotecan and paclitaxel, 10-1000 ng/wipe for docetaxel, and 100-10000 ng/wipe for vindesine, vincristine and vinblastine, respectively.

Table 4 showed the positive ratio of detection and amounts of detected drugs. The surfaces of the BSCs were occasionally contaminated, meanwhile, contaminations in the surfaces of the tables inside a separated area and outside one were at low levels.

Sampling spots	A) Surfaces of the BSCs		B) Surfaces of the tables inside a separated area		C) Surfaces of the tables outside a separated area	
Compounds	Positive samples (n = 28)	Amounts (ng/wipe)	Positive samples (n = 14)	Amounts (ng/wipe)	Positive samples (n = 14)	Amounts (ng/wipe)
Vindesine	0	—	0	—	0	—
Vincristine	0	—	0	—	0	—
Vinblastine	0	—	0	—	0	—
Doxorubicin	5	7 ~ 114	0	—	0	—
Epirubicin	0	—	0	—	0	—
Ifosfamide	0	—	1	7	0	—
Cyclophosphamide	9	5 ~ 68	1	21	1	7
Irinotecan	0	—	0	—	0	—
Docetaxel	3	9 ~ 348	0	—	0	—
Paclitaxel	1	19	1	14	0	—

Table 4. Frequencies and amounts of drugs detected in wipe samples.

4. Discussion

Since Falck's report [Falck et al.,1979], many healthcare workers are concerned about the risk of occupational exposures to hazardous drugs. Previously, in studies of external exposures [Sessink et al.,1992; Minoia et al.,1998; Sabatini et al.,2005; Hedmer et al.,2008] and studies in internal exposures [Hirst et al.,1984; Sessink et al.,1992, 1994, 1995; Ensslin et al.,1994; Burgaz et al.,1999; Pethran et al.,2003], cyclophosphamide was frequently used as a marker of occupational exposures because of its slight volatility and human genotoxicity [Connor et al.,2000], reproductive toxicity [Anderson et al.,1995], carcinogenicity [IARC,1981], and ease of detection [Turci et al.,2002; Barbieri et al.,2006].

We also focused on occupational exposures to several chemotherapeutic drugs including cyclophosphamide, and concluded the precautionary measures we took were adequate to prevent occupational exposures [Ikeda et al.,2007; Maeda et al.,2010].

In this paper, because a progress of chemotherapy requires more versatile monitoring, we developed a multicomponent analysis method of chemotherapeutic drugs, and assessed the levels of environmental contaminations involved in preparation of chemotherapeutic drugs. Anthracycline drugs were light sensitive and were absorbed on glass containers [Lachatre et al.,2000] and vincalkaloid drugs were absorbed on some materials of tubes [Van Tellingen et al.,1991], we selected black Eppendorf tubes for stock solutions and working solutions, and light-blocking polypropylene tubes through extraction.

We prepare chemotherapeutic drugs in three different ways; (i) prepare drugs for outpatients on the BSCs equipped on a separated area, maintained constant a negative pressure, in hospital pharmacy; (ii) prepare drugs for inpatients on the BSCs equipped on a separated area in hospital pharmacy; (iii) prepare drugs for inpatients required special care (such as pretreatment for hematopoietic stem cell transplantation) on the BSCs equipped at the corner of the nurse station in a biological clean ward. We focused on environment of preparation for outpatients, because numbers of drugs were prescribed for outpatients and fast preparations were needed.

Our multicomponent analysis method had sufficient sensitivities and had conveniences enough for regular monitoring. Lower limits of quantitation were 5 ng/wipe for doxorubicin, epirubicin, ifosfamide, cyclophosphamide, irinotecan and paclitaxel, 10 ng/wipe for docetaxel, and 100 ng/wipe for vindesine, vincristine and vinblastine, respectively.

In the environmental assessments, vindesine, vincrisitine, vinblastine, epirubicin and irinotecan were not detected. Although doxorubicin, cyclophosphamide, docetaxel and paclitaxel were occasionally detected in the surfaces of the BSCs, frequencies and amounts of drugs detected were low levels in the surfaces of the working tables inside a separated area and outside one.

Inhalation and dermal penetration were presumed to be the main routes of occupational exposures. Several researchers [McDevitt et al.,1993; Sessink et al.,1994; Minoia et al.,1998] compared the urinary cyclophosphamide excretion levels and concentrations of cyclophosphamide in air during preparation of drugs. They concluded that the amounts of cyclophosphamide inhaled were much lower than amounts of cyclophsphamide excreted in urine, and that inhalation was not the main route of exposures resulting in high levels of cyclophosphamide. Fransman [Fransman et al.,2004, 2005] reported that dermal exposure predominantly occurred on the hands and sporadically on the forehead and forearms, and he mentioned that greater than 90% exposures were prevented by using latex gloves at the time of preparations. These reports showed that dermal exposures occurred on the hands were the main route of internal exposures. Thus, routine monitoring of unexpected environmental contaminations is very useful for preventing occupational exposures.

5. Conclusion

We developed a multicomponent analysis method of ten major chemotherapeutic drugs, and assessed the levels of environmental contaminations involved in preparation of chemotherapeutic drugs. We revealed that environmental contaminations outside BSCs were at low levels, and concluded that adequate precautionary measures and improved awareness regarding handling of chemotherapeutic drugs could reduce the risk of occupational exposures.

6. References

Anderson, D., Bishop, J. B., Garner, R. C., Ostrosky-Wegman, P., & Selby, P. B. (1995). Cyclophosphamide: review of its mutagenicity for an assessment of potential germ cell risks. *Mutation Research*, Vol. 330, No. 1-2, pp. 115-181, ISSN 0027-5107

Barbieri, A., Sabatini, L., Indiveri, P., Bonfiglioli, R., Lodi, V., & Violante, F. S. (2006). Simultaneous determination of low levels of methotrexate and cyclophosphamide in human urine by micro liquid chromatography/electrospray ionization tandem mass spectrometry. *Rapid Communications in Mass Spectrometry*, Vol. 20, No. 12, pp. 1889-1893, ISSN 0951-4198

Burgaz, S., Karahalil, B., Bayrak, P., Taskin, L., Yavuzaslan, F., Bokesoy, I., Anzion, R. B., Bos, R. P., & Platin, N. (1999). Urinary cyclophosphamide excretion and micronuclei frequencies in peripheral lymphocytes and in exfoliated buccal epithelial cells of nurses handling antineoplastics. *Mutation Research*, Vol. 439, No. 1, pp. 97-104, ISSN 0027-5107

Connor, T. H., Shults, M., & Fraser, M. P. (2000). Determination of the vaporization of solutions of mutagenic antineoplastic agents at 23 and 37 degrees C using a desiccator technique. *Mutation Research*, Vol. 470, No. 1, pp. 85-92, ISSN 0027-5107

Ensslin, A. S., Stoll, Y., Pethran, A., Pfaller, A., Rommelt, H., & Fruhmann, G. (1994). Biological monitoring of cyclophosphamide and ifosfamide in urine of hospital personnel occupationally exposed to cytostatic drugs. *Occupational and Environmental Medicine*, Vol. 51, No. 4, pp. 229-233, ISSN 1351-0711

Falck, K., Grohn, P., Sorsa, M., Vainio, H., Heinonen, E., & Holsti, L. R. (1979). Mutagenicity in urine of nurses handling cytostatic drugs. *Lancet*, Vol. 1, No. 8128, pp. 1250-1251, ISSN 0140-6736

Fransman, W., Vermeulen, R., & Kromhout, H. (2004). Occupational dermal exposure to cyclophosphamide in Dutch hospitals: a pilot study. *Annals of Occupational Hygiene*, Vol. 48, No. 3, pp. 237-244, ISSN 0003-4878

Fransman, W., Vermeulen, R., & Kromhout, H. (2005). Dermal exposure to cyclophosphamide in hospitals during preparation, nursing and cleaning activities. *International Archives of Occupational and Environmental Health*, Vol. 78, No. 5, pp. 403-412, ISSN 0340-0131

Goolsby, T. V., & Lombardo, F. A. (2006). Extravasation of chemotherapeutic agents: prevention and treatment. *Seminars in Oncology*, Vol. 33, No. 1, pp. 139-143, ISSN 0093-7754

Hedmer, M., Tinnerberg, H., Axmon, A., & Jonsson, B. A. (2008). Environmental and biological monitoring of antineoplastic drugs in four workplaces in a Swedish hospital. *International Archives of Occupational and Environmental Health*, Vol. 81, No. 7, pp. 899-911, ISSN 0340-0131

Hirst, M., Tse, S., Mills, D. G., Levin, L., & White, D. F. (1984). Occupational exposure to cyclophosphamide. *Lancet*, Vol. 1, No. 8370, pp. 186-188, ISSN 0140-6736

IARC (1981). *Monographs on the Evaluation of the Carcinogenic Risk of Chemicals to Humans: Some Antineoplastic and Immunosuppressive Agents, Cyclophosphamide*, International Agency for Research on Cancer, Lyon, France

Ikeda, K., Yagi, Y., Takegami, M., Lu, Y., Morimoto, K., & Kurokawa, N. (2007). Efforts to ensure safety of hospital pharmacy personnel occupationally exposed to antineoplastic drugs during a preparation task. *Hospital pharmacy (Philadelphia, PA)*, Vol. 42, No. 3, pp. 209-218, ISSN 0018-5787

Jakab, M. G., Major, J., & Tompa, A. (2001). Follow-up genotoxicological monitoring of nurses handling antineoplastic drugs. *Journal of Toxicology and Environmental Health, Part A*, Vol. 62, No. 5, pp. 307-318, ISSN 1528-7394

Kasuba, V., Rozgaj, R., & Garaj-Vrhovac, V. (1999). Analysis of sister chromatid exchange and micronuclei in peripheral blood lymphocytes of nurses handling cytostatic drugs. *Journal of Applied Toxicology*, Vol. 19, No. 6, pp. 401-404, ISSN 0260-437X

Kopjar, N., & Garaj-Vrhovac, V. (2001). Application of the alkaline comet assay in human biomonitoring for genotoxicity: a study on Croatian medical personnel handling antineoplastic drugs. *Mutagenesis*, Vol. 16, No. 1, pp. 71-78, ISSN 0267-8357

Lachatre, F., Marquet, P., Ragot, S., Gaulier, J. M., Cardot, P., & Dupuy, J. L. (2000). Simultaneous determination of four anthracyclines and three metabolites in human serum by liquid chromatography-electrospray mass spectrometry. *Journal of Chromatography B*, Vol. 738, No. 2, pp. 281-291, ISSN 1387-2273

Lanza, A., Robustelli della Cuna, F. S., Zibera, C., Pedrazzoli, P., & Robustelli della Cuna, G. (1999). Somatic mutations at the T-cell antigen receptor in antineoplastic drug-exposed populations: comparison with sister chromatid exchange frequency. *International Archives of Occupational and Environmental Health*, Vol. 72, No. 5, pp. 315-322, ISSN 0340-0131

Larson, R. R., Khazaeli, M. B., & Dillon, H. K. (2003). Development of an HPLC method for simultaneous analysis of five antineoplastic agents. *Applied Occupational and Environmental Hygiene*, Vol. 18, No. 2, pp. 109-119, ISSN 1047-322X

Lissalde, S., Mazzella, N., Fauvelle, V., Delmas, F., Mazellier, P., & Legube, B. (2011). Liquid chromatography coupled with tandem mass spectrometry method for thirty-three pesticides in natural water and comparison of performance between classical solid phase extraction and passive sampling approaches. *Journal of Chromatography A*, Vol. 1218, No. 11, pp. 1492-1502, ISSN 0021-9673

Maeda, S., Miyawaki, K., Matsumoto, S., Oishi, M., Miwa, Y., & Kurokawa, N. (2010). Evaluation of environmental contaminations and occupational exposures involved in preparation of chemotherapeutic drugs. *Yakugaku Zasshi Journal of the Pharmaceutical Society of Japan*, Vol. 130, No. 6, pp. 903-910, ISSN 0031-6903

McDevitt, J. J., Lees, P. S., & McDiarmid, M. A. (1993). Exposure of hospital pharmacists and nurses to antineoplastic agents. *Journal of Occupational Medicine*, Vol. 35, No. 1, pp. 57-60, ISSN 0096-1736

Miao, L., Cai, W., & Shao, X. (2011). Rapid analysis of multicomponent pesticide mixture by GC-MS with the aid of chemometric resolution. *Talanta*, Vol. 83, No. 4, pp. 1247-1253, ISSN 0039-9140

Minoia, C., Turci, R., Sottani, C., Schiavi, A., Perbellini, L., Angeleri, S., Draicchio, F., & Apostoli, P. (1998). Application of high performance liquid chromatography/ tandem mass spectrometry in the environmental and biological monitoring of health care personnel occupationally exposed to cyclophosphamide and ifosfamide. *Rapid Communications in Mass Spectrometry*, Vol. 12, No. 20, pp. 1485-1493, ISSN 0951-4198

Nabeshima, T. S., Toru., Sugiura, S., Tanimura, M., Nakao, M., Nakanishi, H., & Kato, K. (2008). Guidelines of an aseptically preparation and of antineoplastic drugs preparation (in Japanese). *Journal of Japanese Society of Hospital Pharmacists*, Vol. 44, No. 1, pp. 18-20, ISSN 1341-8815

NIOSH (2004). *Preventing Occupational Exposure to Antineoplastic and Other Hazardous Drugs in Health Care Settings*, The National Institute for Occupational Safety and Health, Atlanta, USA

Nussbaumer, S., Fleury-Souverain, S., Antinori, P., Sadeghipour, F., Hochstrasser, D. F., Bonnabry, P., Veuthey, J. L., & Geiser, L. (2010). Simultaneous quantification of ten cytotoxic drugs by a validated LC-ESI-MS/MS method. *Analytical and Bioanalytical Chemistry*, Vol. 398, No. 7-8, pp. 3033-3042, ISSN 1618-2642

Pethran, A., Schierl, R., Hauff, K., Grimm, C. H., Boos, K. S., & Nowak, D. (2003). Uptake of antineoplastic agents in pharmacy and hospital personnel. Part I: monitoring of urinary concentrations. *International Archives of Occupational and Environmental Health*, Vol. 76, No. 1, pp. 5-10, ISSN 0340-0131

Roth, S., Norppa, H., Jarventaus, H., Kyyronen, P., Ahonen, M., Lehtomaki, J., Sainio, H., & Sorsa, M. (1994). Analysis of chromosomal aberrations, sister-chromatid exchanges and micronuclei in peripheral lymphocytes of pharmacists before and after working with cytostatic drugs. *Mutation Research*, Vol. 325, No. 4, pp. 157-162, ISSN 0027-5107

Sabatini, L., Barbieri, A., Tosi, M., & Violante, F. S. (2005). A new high-performance liquid chromatographic/electrospray ionization tandem mass spectrometric method for the simultaneous determination of cyclophosphamide, methotrexate and 5-fluorouracil as markers of surface contamination for occupational exposure monitoring. *Journal of Mass Spectrometry*, Vol. 40, No. 5, pp. 669-674, ISSN 1076-5174

Sebok, A., Vasanits-Zsigrai, A., Helenkar, A., Zaray, G., & Molnar-Perl, I. (2009). Multiresidue analysis of pollutants as their trimethylsilyl derivatives, by gas chromatography-mass spectrometry. *Journal of Chromatography A*, Vol. 1216, No. 12, pp. 2288-2301, ISSN 0021-9673

Sessink, P. J., Boer, K. A., Scheefhals, A. P., Anzion, R. B., & Bos, R. P. (1992). Occupational exposure to antineoplastic agents at several departments in a hospital. Environmental contamination and excretion of cyclophosphamide and ifosfamide in urine of exposed workers. *International Archives of Occupational and Environmental Health*, Vol. 64, No. 2, pp. 105-112, ISSN 0340-0131

Sessink, P. J., Van de Kerkhof, M. C., Anzion, R. B., Noordhoek, J., & Bos, R. P. (1994). Environmental contamination and assessment of exposure to antineoplastic agents by determination of cyclophosphamide in urine of exposed pharmacy technicians:

is skin absorption an important exposure route? *Archives of Environmental Health*, Vol. 49, No. 3, pp. 165-169, ISSN 0003-9896

Sessink, P. J., Kroese, E. D., van Kranen, H. J., & Bos, R. P. (1995). Cancer risk assessment for health care workers occupationally exposed to cyclophosphamide. *International Archives of Occupational and Environmental Health*, Vol. 67, No. 5, pp. 317-323, ISSN 0340-0131

Skoczynska, E., Korytar, P., & De Boer, J. (2008). Maximizing chromatographic information from environmental extracts by GCxGC-ToF-MS. *Environmental Science and Technology*, Vol. 42, No. 17, pp. 6611-6618, ISSN 0013-936X

Sottani, C., Turci, R., Schierl, R., Gaggeri, R., Barbieri, A., Violante, F. S., & Minoia, C. (2007). Simultaneous determination of gemcitabine, taxol, cyclophosphamide and ifosfamide in wipe samples by high-performance liquid chromatography/tandem mass spectrometry: protocol of validation and uncertainty of measurement. *Rapid Communications in Mass Spectrometry*, Vol. 21, No. 7, pp. 1289-1296, ISSN 0951-4198

Sottani, C., Rinaldi, P., Leoni, E., Poggi, G., Teragni, C., Delmonte, A., & Minoia, C. (2008). Simultaneous determination of cyclophosphamide, ifosfamide, doxorubicin, epirubicin and daunorubicin in human urine using high-performance liquid chromatography/electrospray ionization tandem mass spectrometry: bioanalytical method validation. *Rapid Communications in Mass Spectrometry*, Vol. 22, No. 17, pp. 2645-2659, ISSN 0951-4198

Tomioka, K., & Kumagai, S. (2005). Health risks of occupational exposure to anticancer (antineoplastic) drugs in health care workers (in Japanese). *Sangyo Eiseigaku Zasshi*, Vol. 47, No. 5, pp. 195-203, ISSN 1341-0725

Turci, R., Sottani, C., Ronchi, A., & Minoia, C. (2002). Biological monitoring of hospital personnel occupationally exposed to antineoplastic agents. *Toxicology Letters*, Vol. 134, No. 1-3, pp. 57-64, ISSN 0378-4274

Van Tellingen, O., Beijnen, J. H., & Nooyen, W. J. (1991). Analytical methods for the determination of vinca alkaloids in biological specimens: a survey of the literature. *Journal of Pharmaceutical and Biomedical Analysis*, Vol. 9, No. 10-12, pp. 1077-1082, ISSN 0731-7085

Yoshida, J., Tei, G., Mochizuki, C., Masu, Y., Koda, S., & Kumagai, S. (2009). Use of a closed system device to reduce occupational contamination and exposure to antineoplastic drugs in the hospital work environment. *Annals of Occupational Hygiene*, Vol. 53, No. 2, pp. 153-160, ISSN 1475-3162

6

Production of Persistent Organic Pollutants from Cement Plants

Flávio Aparecido Rodrigues
Universidade de Mogi das Cruzes, Laboratório de Materiais e Superfícies (LABMAR)
Brazil

1. Introduction

In 2001, The United Nations Environmental Programme (UNEP), during the Stockholm Convention (UNEP 2001) listed the so-called "dirty dozen" chemicals, the persistent organic pollutants (POPs). The agreement was amended in 2009, including new classes of compounds. This agreement was ratified by 150 countries. Table 1 lists the POPs regulated by UN.

Persistent Organic Pollutants		
Aldrin, Chlordane , DDT, Dieldrin,, Endrin, Chlordecone	Pentachlorobenzene, Hexabromobiphenyl, Hexabromodiphenyl ether and heptabromodiphenyl ether	Polychlorinated dibenzo-*p*-dioxins and dibenzofurans (PCDD/PCDF)
Heptachlor, Hexachlorobenzene (HCB), Mirex, Toxaphene	Tetrabromodiphenyl ether, pentabromodiphenyl ether	Polychlorinated biphenyls (PCB)
Alpha hexachlorocyclohexane, Beta hexachlorocyclohexane, Lindane	Perfluorooctane sulfonic acid and its salts	perfluorooctane sulfonyl fluoride

Table 1. Persistent organic pollutants according to Stockholm convention agreement (2009)

The term persistent organic pollutant was used much before 2011 (Rantio, 1996) in scientific literature. However, it is important to recognize that UN convention was a significant step in order to bring many countries together with a unified goal: elimination and/ or restriction of POPs.

The list presented in Table 1 comprises all chemicals considered persistent organic pollutants. It shows that persistent organic pollutants, in fact, constitute a diverse class of chemical compounds with specific physical-chemical characteristics. On the other hand, it is important to note that there are many important differences among these chemicals. For instance, alpha and beta hexachlorocyclohexane were completely banned and can not be produced under any circumstances, for commercial use. It can be produced however for research purposes. On the other hand, the production of DDT (1,1,1-trichloro-2,2-bis (4-chlorophenyl)ethane) is restricted and its use should be limited. DDT is still used to control malaria in specific regions (Karakus et al., 2006). Other POPs, such as polychlorinated dibenzo-*p*-dioxins and dibenzofurans (PCDD/PCDF) are not intentionally produced; rather, they are generated as a result of combustion processes from industrial activities (Abad et al., 2002). From the chemical viewpoint, there are hundreds of POP's, when considering congeners within a specific family of compounds (Jones & de Voogt, 1999). For instance, the term dioxin includes 75 PCDD and 135 PCDFs (Srogi, 2008). There are many differences between these compounds in terms of toxicity or degradation in the environment; however, it is much more convenient to use the "general" name, rather than describe each component individually.

Historically the large-scale utilization of most POPs grew faster after the World War II, especially as insecticide. DDT is perhaps the most "popular" organic pollutant. Table 2 lists the utilization of DDT in 10 countries, around the world.

Country	Agricultural use (kTon)	Period
USA	590	1947-1972
Former Soviet Union	320	1952-1971
China	260	1952-1983
Mexico	180	1953-2000
Brazil	106	1947-1998
India	75	1947-1990
Egypt	66	1952-1972
Guatemala	60	1947-1985
Italy	46	1948-1987

Table 2. Agricultural use of DDT in history

It is apparent from table 2 that large amounts of DDT were intentionally released in the environment. It should be pointed out that many other insecticides were produced and used in every continent. It is not surprising that UN is acting directly to minimize the harmful effects of these chemicals on the environment. The dissemination of POPs is also due to their large range of applications. For example, aldrin, chlordane, dieldrin and endrin were used as insecticides, hexabromobiphenyl is a flame retardant (Sjödin et al., 2006) and perfluorinated organic compounds are used as insecticide, lubricants and surfactants (Liu et al., 2007). Despite having chemical differences, POPs show common properties, making them harmful to the environment. Probably, the most important aspects associated to POPs are the toxicity, the slow degradation in the environment and their bioaccumulation. These aspects are briefly discussed in the following section.

1.1 Persistent organic pollutants and their toxicity

World Health Organization (WHO) listed the toxicity of several POPs as shown in Table 3. All of the listed organic pollutants are toxic and few of these are extremely lethal, even in very small amounts. For instance, the International Programme on Chemical Safety report (IPCS) presents the toxicity of the dirty dozen, using the concept of lethal doses, LD_{50}. LD_{50} is the concentration of the compound to kill 50% of animal species under evaluation.

Compound	Toxicity
Aldrin	guinea pig and the hamster is 33 and 320 mg/kg bw, respectively.
Chlordane	The acute oral LD_{50} in the rat is 200-590 mg/kg bw
DDT (1,1,1-trichloro-2,2-bis (4-chlorophenyl) ethane)	The acute oral LD_{50} in the rat is 113-118 mg/kg bw
Dieldrin	The acute oral LD_{50} in the mouse and rat ranges from 40 - 70 mg/kg bw
Endrin	The acute oral LD_{50} values are between 3-43 mg/kg bw, for long term toxicity in the rat.
Heptachlor	The acute oral LD_{50} values are between 40-119 mg/kg bw.
Hexachlorobenzene (HCB)	The acute toxicity is low; oral LD_{50} value of 3.5 mg/g bw in the rat has been reported.
Mirex	The acute oral LD_{50} in the rat is 235 mg/kg bw.
Toxaphene	The acute oral toxicity ranges from 49 (dogs) to 365 (guinea pigs) mg/kg bw.
Polychlorinated biphenyls (PCBs)	The acute toxicity is generally low; oral LD_{50} value of 1 g/kg bw in the rat has been reported.
Polychlorinated dibenzo-p-dioxins (PCDDs/Dioxins) and polychlorinated dibenzofurans PCDFs/Furans)	Effects on the immune system in the mouse have been reported at 10 ng/kg

Table 3. Toxicity reported for some persisten organic pollutant (Stockholm Convention on Persistent Organic Pollutants (POPs), 2009)

On the other hand, Schecter et. al (2006) compared the toxicity of dioxins, using toxic equivalent factor (TEQ). In this scale, the number 1 was assigned to the compound 2,3,7,8-tetrachlorodibenzo-p-dioxin (TCDD), the most toxic among dioxins. Table 4 present a list of toxicity of dioxins using TEQ scale. As it can be seen, toxicity is dependent on chemical structure of compounds, and some dioxins present low toxicity.

There is not a comprehensive scale of toxicity, able to correlate all persistent organic pollutants. On the other hand, there are many studies dealing with toxicity of POPs. For instance, polybrominated diphenyl ethers (PBDEs) act directly on the brain (neurotoxic) and may cause behavioral disorders, specially the HO-PBDE form (Dingemans et al., 2011). Lee et al., (2010) suggested that some POPs may be associated to type 2 Diabetes. Many other harmful aspects of POPs on human beings, are presented in literature, such as tumour promoter (Shin et al., 2010), pancreatic cancer (Hardell et al., 2007), cardiovascular diseases (Lind et al., 2004) and neurological problems (Weiss, 2011).

1.2 Degradation in the environment

Many POPs were initially used as insecticide and deliberately spread over soils, rivers and air. Their use started in 1940 and for many years, thousands of tons were freely dispersed in

Compound class	isomer	TEF
Dioxins	2,3,7,8-Tetra-CDD	1
	1,2,3,7,8-Penta-CDD	1
	1,2,3,4,7,8-Hexa-CDD	0.1
	1,2,3,6,7,8-Hexa-CDD	0.1
	1,2,3,7,8,9-Hexa-CDD1	0.1
	1,2,3,4,6,7,8-Hepta-CDD	0.01
	OCDD	0,0001
Dibenzofurans	2,3,7,8-Tetra-CDF	0.1
	1,2,3,7,8-Penta-CDF	0.05
	2,3,4,7,8-Penta-CDF	0.5
	1,2,3,4,7,8-Hexa-CDF	0.1
	1,2,3,6,7,8-Hexa-CDF	0.1
	1,2,3,7,8,9-Hexa-CDF	0.1
	2,3,4,6,7,8-Hexa-CDF	0.1
	1,2,3,4,6,7,8-Hepta-CDF	0.01
	1,2,3,4,7,8,9-Hepta-CDF	0.01
	OCDF	0.0001
Coplanar PCBs	3,30,4,40-TCB (77)	0.0001
	3,4,40,5-TCB (81)	0.0001
	3,30,4,40,5-PeCB (126)	0.1
	3,30,4,40,5,50-HxCB (169)	0.01
Mono-ortho-PCBs	2,3,30,4,40-PeCB (105)	0.0001
	2,3,4,40,5-PeCB (114)	0.0005
	2,30,4,40,5-PeCB (118)	0.0001
	20,3,4,40,5-PeCB (123)	0.0001
	2,3,30,4,40,5-HxCB (156)	0.0005
	2,3,30,4,40,50-HxCB (157)	0.0005
	2,30,4,40,5,50-HxCB (167)	0.00001
	2,3,30,4,40,5,50-HpCB (189)	0.0001

Table 4. Comparative toxicity of dioxins. Toxic Equivalent Factor (TEQ) = 1 for compound 2,3,7,8-tetrachlorodibenzo-p-dioxin, the most toxic substance

many parts of the world. Of course, at that time, there was no complete evidence that these chemicals could be so harmful to the environment. Nowadays, even considering restrictions or banning of POPs they can still be found in different regions. There are indications that some POPs are still in use (Minh et al., 2007). The understanding of the behavior of a chemical in the environment is in fact, a very complex task. For instance, when a pesticide is spread over an agricultural area many phenomena may have place. Part of the insecticide may be simply volatized, remaining in the air. In this case, it can be immediately spread over a much larger area. In this case, atmospheric conditions will be important, such as wind intensity, the occurrence of rain, and even local temperature may dictate the fate of these components. When the pesticide contacts soil, it may interact with organic material and be held for an indefinite period of time. It can be degraded or remain stable for long periods. Also, it can be leached and reach rivers or the sea. A particular threat in the case of

POPs is the relatively slow degradation process. This is also a multifaceted development. Chemical degradation of POPs can take years or even decades to be completed. For example, technical chlordane (a mixture of more than 140 components) may persist in soil for more than 20 years (Mattina et. al, 1999). Lindane was found in many areas of France, even in regions without agricultural use (Villanneau et al., 2009). Similar observations and results can be found in the literature (Carlson et al., 2010), dealing with diverse situations, such as latitude and population, all around the world (Liu et. Al, 2009; Kurt-Karakus et al., 2006; Shuthirasingh, et. al., 2010). This behavior is observed under different environmental conditions, such as water, atmosphere or soils (Ramos et. al., 2001; Schwarzenbach et al., 2010; Pozo et al., 2011). Therefore, this degradation kinetic poses an additional and important threat. These substances can be spread over very long distances (Brzuzy and Hites, 1996). Also, the transport mechanism can be very complex and many variables are involved (Mackay and Arnot, 2011). Perhaps the best way to understand the risks posed by POPs is to consider Polar Regions, Antarctic and Artic. At first glance, these areas could be considered remote and free from deleterious human activities. Unfortunately, this is not the case. Contamination in Antarctic continent was first reported in 1966 (Corsolini, 2009). Many POPs are relatively volatile under environmental conditions. These chemical may present a liquid-gas transition and be transported by the wind, according to seasonal conditions. This process (phase transition) may occur several times since the degradation is very slow. More important, when POPs reach cold climate like in Polar Regions, they tend to be trapped, due to very low temperature. For instance, there are many researches dealing with contamination in Arctic and Antarctic regions (Bargagli, 2008; Tin et al., 2009). (Wania, 2003). In this sense, even unintentional production of POPs is not a localized problem; rather, it should be treated as a global concern. It is very usual to say that POPs are subject to long-range atmospheric transport (Shuntirasingh et al., 2010).

1.3 Bioaccumulaton

From the chemical viewpoint, all these organic pollutants are very hydrophobic and these chemicals will minimize contact with water, tending to migrate to organic matter. For example, when these chemical reach soils, they will be preferentially adsorb on bioavailable organic components, such as humic substances (Wu et al., 2011). Also, they can be absorbed by plants and living organisms. This is an example of how POPs may enter in the food chain and will be assimilated by animals, for example. They will preferentially migrate to fatty tissues. For instance, a recent study (Kim et al., 2010) indicates that the concentration of POP´s is much higher in obese individuals. Once again, since the degradation is very slow, they will be accumulated over the years. In a similar way, POPs are easily transported by atmosphere. When in contact with water they also will be transported to planktonic organisms, entering the food chain. POPs were found as contaminants in birds (Chen and Hale, 2010; Vorkamp et al., 2009), penguins (Corsoloni et al., 20060, fish (Ondarza et al., 2011; Hardell et al., 2010), sea lions (Alava et al., 2011), among others.

2. Unintentional production of POPs

From the perspective presented so far, it seems that POPs should be banned from any industrial or commercial activity, in order to preserve the environment. However this is not a simple task. Many POPs are unintentionally produced; they are, in fact, a result of industrial activities. These chemical, polychlorinated dibenzo-*p*-dioxins and dibenzofurans

(PCDD/PCDF), polychlorinated biphenyls (PCB), can be released into atmosphere as a sub-product. The following is the original text issued by UNEP.

"Hexachlorobenzene, pentachlorobenzene, polychlorinated biphenyls, and polychlorinated dibenzo-p-dioxins and dibenzofurans are unintentionally formed and released from thermal processes involving organic matter and chlorine as a result of incomplete combustion or chemical reactions. The following industrial source categories have the potential for comparatively high formation and release of these chemicals to the environment:

a. Waste incinerators, including co-incinerators of municipal, hazardous or medical waste or of sewage sludge;
b. Cement kilns firing hazardous waste;
c. Production of pulp using elemental chlorine or chemicals generating elemental chlorine for bleaching;
d. The following thermal processes in the metallurgical industry:
i.   Secondary copper production;
ii. Sinter plants in the iron and steel industry;
iii. Secondary aluminium production;
iv. Secondary zinc production.

Hexachlorobenzene, pentachlorobenzene, polychlorinated biphenyls, and polychlorinated dibenzo-p-dioxins and dibenzofurans may also be unintentionally formed and released from the following source categories, including:

a. Open burning of waste, including burning of landfi ll sites;
b. Thermal processes in the metallurgical industry not mentioned in Part II;
c. Residential combustion sources;
d. Fossil fuel-fi red utility and industrial boilers;
e. Firing installations for wood and other biomass fuels;
f. Specific chemical production processes releasing unintentionally formed persistent organic pollutants, especially production of chlorophenols and chloranil;
g. Crematoria;
h. Motor vehicles, particularly those burning leaded gasoline;
i. Destruction of animal carcasses;
j. Textile and leather dyeing (with chloranil) and finishing (with alkaline extraction);
k. Shredder plants for the treatment of end of life vehicles;
l. Smouldering of copper cables;
m. Waste oil refineries"

As it can be seen, there are many different activities able to unintentionally generate some POPs. Of course, some activities do not generate significant amounts of persistent organic pollutants, while others, due to massive production, may pose a high risk to environment. Here we will focus only on cement production. This industrial area may be very useful to represent the challenges in our modern society, regarding economic needs and environmental concern.

3. Cement production: Background and economic aspects

Cement is produced in almost all countries in the world. Concrete is a composite material formed by cement, sand, rocks and water. There are many kinds of cements and concretes.

Concrete is versatile material. It can be used under many environmental conditions, such as cold and hot climate, in the sea, under very high pressure and temperature. According to Meyer (2009) the world production of concrete is superior to 10 billion tons yearly. Table 5 lists the top-10 countries in cement production, in the period between 2004-2007, as well as world total manufacture. (US Department of Interior, 2005-2009; Rodrigues and Joekes, 2011).

Country	2004	2005	2006	2007
China	934,000	1,040,000	1,200,000	1,350,000
India	125,000*	145,000*	155,000*	170,000*
USA	99,000	101,000	99,700	96,500
Japan	67,400	69,600	69,900	67,700
Russia	43,000*	48,700	54,700	59,900
Korea	53,900	51,400	55,000	57,000
Spain	46,800	50,300	54,000e	54,500
Turkey	38,000	42,800	47,500	49,500
Italy	38,000*	46,400	43,200	47,500
Brazil	38,000*	36,700	39,500	46,400
World	2,130,000*	2,310,000*	2,550,000*	2,770,000*

* estimative

Table 5. Cement production for different countries during the years 2004-2007

In a recent review (Rodrigues and Joekes, 2011) it was presented some implications and consequences of cement production. As it can be seen in Table 5, China produces about 50% of cement in the world. At a first glance, this massive production could be related to Chinese economic growth. Although this is a correct assumption, it does not explain the current situation in that country. There are many important aspects to be considered (Shen et al., 2005; Zhang and Zhao, 2000). China has experienced an unprecedented urbanization. It is estimated that urban population reached 456 million in 2000. In 1978, this population was 170 million. It is obvious that urbanization caused many different problems (He et al., 2008). However, in the context presented here, it is important to note that urbanization is the driving-force to production and consume of cement and concrete. In fact, the consumption of these materials is due to the lack of infrastructure. Concrete is used in sanitation, construction of buildings, roads, power plants and so on. Also, along with economic growth, people demand better living standards. Furthermore the mobility of population requires the construction of many kinds of new buildings (Shen et al., 2005).

It seems clear that cement will continue to play an important role in modern society. However, from the environmental viewpoint, cement production can be considered harmful, when the whole process is considered. For instance, for each ton of cement produced, it will render about 1 ton of carbon dioxide. (Worrel et al., 2001). It corresponds to about 5-6% of total CO_2 produced by man. Under the same conditions, it will consume around 60-130 kg of fossil fuel and 110 kWh (Morsli et al., 2007). In USA, cement production consume 0,6% of total energy produced in that country. Also, 1 ton of cement, demands 1.4-1.6 ton of raw-materials. (Horvath, 2004).

In order to understand the generation of POPs from cement manufacture is convenient to describe cement composition and the manufacture process. Table 6 lists the major components of Portland cement and their proportion. It is also important to note, that chemical cement composition is variable, depending on many variables (Rodrigues & Joekes, 2011).

Componente	Formula	% (average)
Tricalcium silicate	Ca_3SiO_5	45-60
Dicalcium silicate	β-Ca_2SiO_4	15-30
Calcium aluminate	$Ca_3Al_2O_6$	2-15
Calcium ferroaluminate	$4CaO.Al_2O_3.Fe_2O_3$	5-15
gypsum	$CaSO_4.1/2H_2O$	5

Table 6. Major components of ordinary Portland cement type I

Calcium silicates are the most important components of Portland cement, although calcium aluminate, calcium ferroaluminate and gypsum are important in the hardening process. Usually, the synthesis of cement is accomplished at around 1500°C. Comparatively, cement is a low-cost material. Many efforts are conducted by cement industry to be competitive, such as location, reduction of energy consumption and distribution (Kendall et al., 2008; Newmark, 1998). Figure 1 presents a simplified diagram for the production of commercial Portland cement.

Cement production is far more complex than presented in figure 1. Specially, the heating of raw-materials may be accomplished in different ways. In order to better understand POPs emissions, two stages are presented here: primary and secondary heating. This division was adopted to simplify the discussion, although this is not the usual method to describe cement production.

The Stockholm convention on persistent organic pollutants (2001) listed the 4 major sources of unintentional production of POP's: waste incinerators, production of pulp using elemental chlorine or chemicals, thermal processes in metallurgical industry and cement kilns.

4. Persistent organic pollutants from cement plants

Cement manufacture demands the use of very high temperatures. It means that energy is a relevant cost in this industry. Several attempts were made through the years to minimize energy consumption. Also, in terms of sustainability, the reduction of fossil or non-renewable fuels consumption is a very attractive option. In many plants, coal is the preferred fuel. European Union countries consume about the energy equivalent to 27 Mt of coal (Kookos et al., 2011) to produce cement.

Perhaps the most interesting approach to reduce the use of fossil fuels is the utilization of co-firing wastes to replace coal. Initially, from the environmental point of view this solution is very interesting, and many residues can be used, such as tires (Prisciandaro, 2003; Carrasco et al., 2002), sewage sludge (Gálves et al., 2007), carpet (Lemieux et al., 2004), among many others. For example, European Union countries dispose approximately 600.000 tons of waste tires each year (Aiello et al., 2009). Table 7 shows several alternative materials

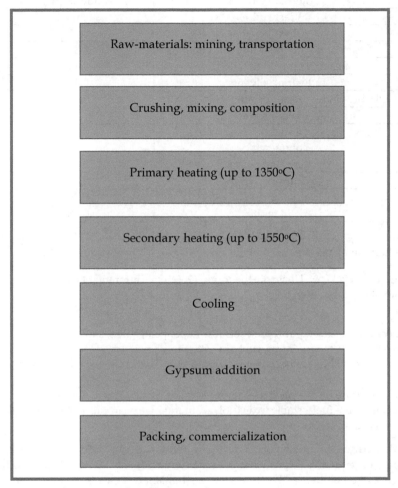

Fig. 1. Major steps of commercial production of cement

that can be incorporated into cement production. As it can be seen, cement production can also be an environmentally-friend process.

According to Cembureau (The European Cement Association) alternative fuels are classified in five categories:

Class 1: gaseous alternative fuels (examples: refinery waste gas, landfill gas),
Class 2: liquid alternative fuels (low chlorine spent solvents, hydraulic oils)
Class 3: pulverized, granulated or finely-crushed solid alternative fuels (sawdust, dried sewage sludge, granulated plastic, animal flours, fine crushed tires),
Class 4: coarse-crushed solid alternative fuels (crushed tires, rubber/ plastic waste, wood waste, reagglomerated organic matter)
Class 5: lump alternative fuels (whole tyres, plastic bales).

Alternative fuel	Amount (Mt)
Waste oil	0.511
Textiles	0.01
Used tires – tire derived fuel (TDF)	0.81
Residue derived fuel (RDF)	0.74
Industrial solvents	0.66
Plastic – industrial and commercial waste	0.46
Meat and bone meal	1.28
Wood, paper, cardboard: industrial and commercial	0.30
Sewage sludge (SS)	0.26
Agricultural waste	0.07
Industrial sludge	0.25
Solid waste	0.45
Oil and oily waste	0.51

Table 7. Alternative materials used by European Union (27) countries in 2005

According Mokrzycki & Uliasz-Bochenczyk (2003) a different classification of fuels to cement industry is presented below:

Solid fuels:
vegetable compounds or natural products (oil shale, peat, barks, sawdust, etc.),
   synthetic products (used tires, rubber waste, waste plastics, etc.),
   others (parts of shredded cars, fuels derived from rejects, household garbage, etc.).
Liquid fuels:
   liquid substitute fuels—easily decomposed, slightly toxic (acid tar, oil residues, etc.),
   liquid substitute fuels, stable toxic (polyaromatic hydrocarbons (PAH), polichlorinated biphenyl (PCB), etc.).

On the other hand, the utilization of so many wastes or alternative materials may bring undesirable consequences. Here we are concerned with the unintentional production of persistent organic pollutants. The formation mechanism of POPs is certainly very complex, since many compounds are possible. For instance, dioxins, PCDD/ PCDFs were studied in the process of the burning of municipal solid waste (Lee et al., 2008). Some reactions involved in this process may be depicted.

The formation of these chemicals from cement kilns occurs by two major mechanisms: by the presence of precursor or *de novo synthesis*. In the precursor mechanism there will be the polycondensation of precursor. Usually these gaseous reactions will take place in the range between 300-600°C. In the *de novo synthesis*, carbon must be present at the solid phase; on the other hand, the presence of oxygen in the gas phase is also fundamental (Fujimori et al., 2010). Usually these reactions take place at 200-400°C.

A possible example of precursor synthesis may be described as follows, considering the utilization of aromatic solvents.

The conversion of HCl to chlorine, in the presence of a catalyst ($CuCl_2$)

$$2CuCl_2 + \frac{1}{2}O_2 \rightarrow Cu_2OCl_2 \tag{1}$$

$$Cu_2OCl_2 + 2HCl \rightarrow 2CuCl_2 + Cl_2 + H_2O \tag{2}$$

Followed by direct transfer of chlorine by a ligand transfer oxidation mechanism

$$ArH + CuCl_2 \rightarrow ArHCl^* + CuCl \tag{3}$$

$$ArHCl^* + CuCl_2 \rightarrow ArCl + CuCl + HCl \tag{4}$$

As the reaction proceeds, will occur a sequential chlorination. The formation of PCDDs, PCDFs and PCBs, PAHs may occur during cement fabrication (van Loo, 2008).

It should be pointed out that engineering aspects such as the kind of kiln, the way how chemicals are used, rate, oxygen levels, etc, may inhibit or minimize the formation of pollutants (Prisciandaro, 2003).

5. Conclusions

Cement production shows interesting aspects regarding the balance between an industrial (pollutant) activity and environmental concerns. The unintentional production of POPs is, certainly, an important issue to deal with. However, it is impossible to imagine a limitation on cement manufacture, based on this consideration. Rather, cement industry and other organization, such as United Nations and governments, should work together to understand and minimize this (and other) problems.

Cement industry shows a high potential to be sustainable. The utilization of residues and wastes from other industrial activities, in fact, should be stimulated, since tons of hazardous components may be eliminated. The generation of persistent organic pollutants, a side-effect, needs a better comprehension. It is very possible that in few years cements plants will be able to operate in order to avoid POPs emissions.

6. Acknowledgements

Author wishes to thank Fundação de Amparo à Pesquisa (FAEP), Universidade de Mogi das Cruzes (UMC), Fundação de Amparo à Pesquisa do Estado de São Paulo (FAPESP) and Financiadora de Estudos e Projetos (FINEP).

7. References

Abad, E.; Caixach, J.; Rivera, J.; Gustems, L.; Massagu, L.; Puig, O. (2002). Surveillance programme on dioxin levels in ambient air sites in Catalonia (Spain).. *Chemosphere*, Vol. 49, No. 7 (November 2002), pp. 697–702 ISSN 0045-6535
Aiello MA, Leuzzi F, Centonze G, Maffezzoli A. (2009). Use of steel fibers recovered from waste tyres as reinforcement in concrete: pull-out behavior, compressive and

flexural strength. *Waste Management,* Vol. 29, No. 6 (June 2009), pp.:1960–1970 ISSN 0956-053X

Al-Assadi, G.; Casati, M. J.; Fernández, J.; Gálvez, J. C. (2011). Effect of the curing conditions of concrete on the behaviour under freeze–thaw cycles. *Fatigue & Fracture of Engineering Materials & Structures,* Vol. 34 No. 5 (July 2011), pp 461-460 ISSN 1460-2695

Alava, J. J.; Ross, P. S.; Ikonomu, M. G.; Cruz, M.; Uzcátegui, G. J.; Dubetz, C.; Salazar, S.; Costa, D. P.; Amtmann, S. V.; Howorth, P.; Gobas, F. A. P.C. (20110). DDT in endangered Galapagos sea lions (*Zalophus wollebaeki*). *Marine Pollution Bulletin,* Vol. 62, No. 4, (April 2011), pp. 660-671ISSN 0025-326X

Bargagli, V. (2008). Environmental contamination in Antarctic ecosystems. *Science of the Total Environment,* Vol. 400, No. 3, (march 2008), pp. 212-226, ISSN 0048-9697

Brzuzy, R.P.; Hites, L.P. (1996). Global Mass Balance for Polychlorinated Dibenzo-*p*-dioxins and Dibenzofurans. *Environmental Science Technology,* Vol.30, No. 6 (June 1996), pp. 1797-1804 ISSN 0013-936X

Carlson, D.; Devault, D.; Swackhamer, D. (2010). On the Rate of Decline of Persistent Organic Contaminants in Lake Trout (*Salvelinus namaycush*) from the Great Lakes, 1970-2003. *Environmental Science. Technology,* Vol 44, No. 6, (March 2010) pp 2004–2010 ISNN 0013-936X

Carrasco F, Bredin N, Heitz M (2002). Gaseous Contaminant Emissions as Affected by Burning Scrap Tires in Cement Manufacturing. Journal Environmental Quality. Vol. 31, No. 5 (September-October 2002) pp.:1484–1490 ISSN 0047-2425

Chen, D.; Hale, R. C. (2010). A global review of polybrominated diphenyl ether flame retardant contamination in birds. *Environment International* Vol. 36, No. 7 (October 2010) pp 800–811 ISSN 0160-4120

Corsolini, S. (2009). Industrial contaminants in Antarctic biota. *Journal of Chromatography A,* Vol. 1216, No. 3 (January 2009), pp. 598-612. *ISSN: 0021-9673*

Corsoloni, S.; Covaci, A.; Ademollo, N.; Focardi, S.; Schepens, P. (2006). Occurrence of organochlorine pesticides (OCPs) and their enantiomeric signatures, and concentrations of polybrominated diphenyl ethers (PBDEs) in the Adélie penguin food web, Antarctica . *EnvironmentalPollution,* Vol. 140, No.2, (September 2006), pp.371-382. ISSN 0269-7491

Dingemans, M.M.L; van den Berg, M.; Westerink, R.H.S. (2011). Neurotoxicity of Brominated Flame Retardants: (In)direct Effects of Parent and Hydroxylated Polybrominated Diphenyl Ethers on the (Developing) Nervous System. *Environmental Health Perspectives* Vol.11, No. 7 (July, 2011) ISSN 0091-676

Fujimori, T.; Takaoka, M.; Morisawa, S. S. (2010) Chlorinated Aromatic Compounds in a Thermal Process Promoted by Oxychlorination of Ferric Chloride. Environmental Science and Technology Vol. 44, No. 6 (March 2010) pp.1974-1979 ISNN 0013-936X

Gálvez, A.; Conesa, J. A.; Martín-Gullón, I.; Font, R. (2007). Interaction between pollutants produced in sewage sludge combustion and cement raw material. *Chemosphere* 69 (2007) 387–394 ISNN 0045-6535

Hardell, L.; Carlberg, M.; Hardell, K.; Björnfoth , H.; Wickbom , G.; Ionescu, M.; van Bavel, B.; Lindström, G. (2007). Decreased survival in pancreatic cancer patients with high concentrations of organochlorines in adipose tissue. *Biomedicine & Pharmacotherapy* Vol. 61, No. 10 (December 2007), pp. 659-664 ISSN *0753-3322*

Hardell, S.; Tilander, H.; Welfinger-Smith, G.; Burger, J.; Carpenter, D. O. (2010). Levels of Polychlorinated Biphenyls (PCBs) and Three Organochlorine Pesticides in Fish from the Aleutian Islands of Alaska. Plos One, Vol. 5, No. 8 (August 2010), ISSN 1932-6203

He, S.; Liu, Y.; Wu, F.; Webster, C. (2008). Poverty incidence and concentration in different social groups in urban China, a case study of Nanjing. *Cities,* Vol. 25, No. 3 (June 2008), pp. 121–132 ISSN 0264-2751

Horvath A (2004). Construction materials and the environment. *Annual Review Environmental Resource,* Vol. 29 pp 181–204 ISSN 1543-5938

Jones, K. C.; de Voogt, P. (1999). Persistent organic pollutants (POPs): state of the science, *Environmental Pollution,* Vol. 100, No.1-3 (January 1999), pp. 209-221, ISSN 0269-7491

Kim,M. J.; Marchand, P.; Henegar, C.; Antignac, J. P.; Alili, R.; Poitou, C.; Bouillot, J. L.; Basdevant, A.; Bizec, B. L.; Barouki, R.; Clément, K. (2010). Fate and Complex Pathogenic Effects of Dioxins and Polychlorinated Biphenyls in Obese Subjects before and after Drastic Weight Loss. *Environmental Health Perspectives* Vol. 119, No. 3 (March 2011) pp. 377-383, ISSN 0091-676

Kookos, I. K.; Pontikes, Y.; Angelopoulos, G. N.; Lyberatos, G. (2011). Classical and alternative fuel mix optimization in cement production using mathematical programming. *Fuel,* Vol. 90, No. 3 (March 2011), pp. 1277–1284 ISSN 0016-2361

Kurt-Karakus, P. B.; Bidleman, T.; Staebler, R.; Jones, K. (2006). Measurement of DDT Fluxes from a Historically Treated Agricultural Soil in Canada. *Environmental Science. Technology,* Vol. 40, No. 15, (August 2006) pp 4578-4585 ISSN ISSN 0013-936X

Lea FM (1971). Chemistry of Cement and Concrete, Chemical Publishing Company, New York, 1971

Lee, D.H.; Steffes, M. W.; Sjödin, A.; Jones, R. S.; Needham, L. L.; Jacobs, D.R. (2010). Low Dose of Some Persistent Organic Pollutants Predicts Type 2 Diabetes: A Nested Case–Control Study. *Environmental Health Perspectives,* Vol. 188, No. 9 (September 2010) pp 1235-1242 ISSN 0091-676

Lee, V. K. C.; Cheung, W-H.; McKay, G. (2007). PCDD/PCDF reduction by the co-combustion process. *Chemosphere* 70, 4, jan (2008) 682–688 ISNN 0045-6535

Lemieux P, Stewart E, Realff M, Mulholland JA (2004). Emissions study of co-firing waste carpet in a rotary kiln. *Journal of Environmental Management,* Vol. 70, No. 1 (January 2004) pp. 27–33 ISSN 0301-4797

Lind, P. M.; Örberg, J.; Edlund, U. B.; Sjöblom, L. Lind, L. (2004). The dioxin-like pollutant PCB 126 (3,3_,4,4_,5-pentachlorobiphenyl) affects risk factors for cardiovascular disease in female rats (2004) *Toxicology Letters* Vol. 150, No. 3, (May 2004) pp 293–299 ISSN *0378-4274*

Liu, C.; Yu, K.; Shi, X.; Wang, J.; Lam, P.K.S.; Wu, R.R.S.; Zhou, B. (2007). Induction of oxidative stress and apoptosis by PFOS and PFOA in primary cultured hepatocytes of freshwater tilapia (*Oreochromis niloticus*). *Aquatic Toxicology,* Vol. 82, No. 2 (May 2007), pp.135–143 ISSN 0166-445X

Liu, Z.; Chen, S.; Quan, X.; Yang, F. (2009). Evaluating the fate of three HCHs in the typically agricultural environment of Liaoning Province, China. *Chemosphere*, Vol. 76, No. 6 (August 2009) pp. 792–798 ISNN 0045-6535

Mackay, D.; Arnot, J. A. (2011). The application of fugacity and activity to simulating environmental fate of organic contaminants, *Journal Chemical Engineering Data*, Vol. 56, No. 4, (April 2011) pp1345-1355 ISSN 1520-5134

Mattina, M. I.; Iannucci-Berger, W.; Dykas, L.; Pardus, J. (1999). Impact of long-term weathering, mobility, and land use on chlordane residues in soil, *Environmental Science Technology*, Vol.3, No pp 2425-2431. (July 1999) ISSN 0013-936X

Meyer C (2009). The greening of the concrete industry. *Cement Concrete Composites* Vol. 31, No. 8 (Setember 2009), pp. 601–605 ISN 0958-9465

Minh, N. H.; Minh, T. B.; Kajiwara, N.; Kunisue, T.; Iwata, H.; Viet, P. H.; Tu, N. P. C. C.; Tuyen, B. C.; Tanabe, S (2007). Pollution sources and occurrences of selected persistent organic pollutants (POPs) in sediments of the Mekong River delta, South Vietnam. *Chemosphere*, Vol. 67, No. 9 (April 2007) 1794–1801 ISSN 0045-6535

Newmark CM (1998). Price and seller concentration in cement: effective oligopoly or misspecified transportation cost? *Economics Letters*, Vol. 60, No. 2, (August 1998), pp. 243–250 ISSN 0165-1765

Ondarza, P. M.; Gonzales, M.; Fillmann, G.; Miglioranza, K. S. B. (2011). Polybrominated diphenyl ethers and organochlorine compound levels in brown trout (*Salmo trutta*) from Andean Patagonia, Argentina. *Chemosphere*, Vol. 83, No. 11 (June, 2011) pp 1597-1602 ISSN 0045-6535

Pozo, K.; Harner, T.; Lee, S. C.; Sinha, R.K.; Sengupta, B.; Loewen, M.; Geethalakshmi, V.; Kannan, K.; Volpi, V. (2011). Assessing seasonal and spatial trends of persistent organic pollutants (POPs) in Indian agricultural regions using PUF disk passive air samplers, *Environmental Pollution*, Vol.159, No.2, (February 2011) pp.646-653, ISSN 0269-7491

Prisciandaro M, Mazziotti G, Veglio F (2003). Effect of burning supplementary waste fuels on the pollutant emissions by cement plants: a statistical analysis of process data. *Resource Conservation Recycling* Vol. 39, No. 2 (September 2003) pp:161-184 ISSN 0921-3449

Ramos, J.; Gavila, A.; Romero, T.; Ize, I. (2011). Mexican experience in local, regional and global actions for lindane elimination, *Environmental Science Police*, Vol.14, No. 5 (May, 2011), pp.503-5099 ISSN 1462-9011

Rantio, T (1996). Chlorohydrocarbons in pulp mill effluent and the environment – III* persistent chlorofydrocarbon pollutants. *Chemosphere*, Vol. 32, No. 2, (January 1996) pp. 253-265, ISSN 0045-6535

Rodrigues, F. A.; Joekes, I. (1994). Water reducing agents of low-molecular weight-supression of air entrapment and slump loss by addition of an organic solvent. *Cement Concrete Research*, vol. 24, No. 5, (May 1994) pp-987-992 ISSN 0008-8846

Rodrigues, F. A.; Joekes, I. (2011). Cement industry: sustainability, challenges and perspectives. *Environmental Chemistry Letters*, Vol 9, No.2 (June 2011) ISSN 1610-3653

Schwarzenbach, R.P.; Egli, T.; Hofstetter, T.B.; Gunten, U.V.; Wehrli, B. (2010). Global Water Pollution and Human Health, *Annual Review Environmental Resource*, Vol.35, No. 4. pp.35:109–136. ISSN 1543-5938

Shen L, Cheng S, Gunson AJ, Wan H (2005). Urbanization, sustainability and the utilization of energy and mineral resources in China. *Cities* Vol. 22, No. 4, (August 2005), pp.287–302 ISSN 0264-2751

Shin, J. Y.; Choi, Y. Y.; Jeon, H. S.; Hwang, J. H.; Kim, S. A.; Kang, J. H.; Chang, Y. S.; Jacobs, D. R.; Park, J. Y.; Lee, D. H. (2010). Low-dose persistent organic pollutants increased telomere length in peripheral leukocytes of healthy Koreans. *Mutagenesis*, Vol. 25, No. 5 (September 2010), pp. 511–516 ISSN 1520-6866

Shuntirasingh, C.; Mmereki, B., Oyiliagu, C.; Lei, Y.; Wania, F. (2010). Fate of Pesticides in the Arid Subtropics, Botswana, Southern Africa. *Environmental Science. Technology*. Vol. 44, No. 21 (November 2010) pp 8082–8088 ISSN 0013-936X ISNN 0045-6535

Sjödin, A.;. Jones, R.S.; Focant, J. F.; Lapeza, C.; Wang,R. Y.; McGahee III, E. E.; Zhang, Y.; Turner, W. E.; Slazyk, B.; Needham, L.L.;. Patterson Jr, D. G. (2006). Retrospective Time-Trend Study of Polybrominated Diphenyl Ether and Polybrominated and Polychlorinated Biphenyl Levels in Human Serum from the United States. *Environmental Health Perspectives*, Vol. 112, No. 6(May 2004) ISSN 0091-6765

Srogi, K. (2008). Levels and congener distributions of PCDDs, PCDFs and dioxin-like PCBs in environmental and human samples: a review. *Environmental Chemistry Letters*, Vol. 6, No. 1 (February 2008), pp.1–28 ISSN 1610-3653

Stockholm Convention on Persistent Organic Pollutants (POPs), Oct. 2001. Interim Secretariat for the Stockholm Convention. United Nations Environment Programme (UNEP) Chemicals, Geneva, Switzerland. www.pops.int, 50 pp.

Su, N.; Peng, Y-N (2001). The characteristics and engineering properties of dry-mix/steam-injection concrete. *Cement and Concrete Research*, Vol. 31 pp No. 4 (April 2001), pp. 609-619 ISSN 0008-8846

Tin, T.; Fleming, Z. L.; Hughes, K.A.; Ainley, D.G.; Convey, P.; Moreno, C.A.; Pfeiffer, S. (2009). Impacts of local human activities on the Antarctic environment, *Antarctic Science*, Vol. 21, No. 1, (February 2009), pp 3–33 ISSN 0954-1020

Van Loo W (2008). Dioxin/furan formation and release in the cement industry. *Environmental Toxicology Pharmacology* Vol. 25, No. 2 (March 2008) pp 128–130 ISSN 1382-6689

Villanneau, E.; Saby, N.P.A.; Arrouays, D.; Jolivet, C. C.; Boulonne, L.; Caria, G.; Barriuso, E.; Bispo, A.; Briand, O. (2009). Spatial distribution of lindane in topsoil of Northern France. Chemosphere 77 (2009), pp. 1249–1255 ISSN 0045-6535

Vorkamp, K.; Thomsen, M.; Moller, S.; Falk, K.; Sorensen, P. B. Persistent organochlorine compounds in peregrine falcon (*Falco peregrinus*) eggs from South Greenland: Levels and temporal changes between 1986 and 2003 (2009). *Environmental International*, Vol. 35, No. 2 pp 336-341 ISSN 0160-4120

Zhang, L; Zhao, SXB (2000). The intersectoral terms of trade and their impact on urbanization in China Post-Communist. *Economies* Vol. 12, No. 4 (December 200), pp. 445-462 ISSN 1746-1049

Wania, F. (2003). Assessing the Potential of Persistent Organic Chemicals for Long-Range Transport and Accumulation in Polar Regions, *Environmental Science Technology*, Vol.37, No., pp 1344-1351(April 2003) ISSN 0013-936X

Weiss, B. (2007). Endocrine disruptors as a threat to neurological function. *Journal of the Neurological Sciences* Vol. 305, No. 1-2 (June 2007) pp. 11–21 ISSN 0022-510X

Wu, P.; Tang, Y.; Wang, W.; Zhu, N.; Li, P.; Wu, J.; Dang, Z.; Wang, X. (2011). Effect of dissolved organic matter from Guangzhou landfill leachate on sorption of phenanthrene by Montmorillonite. *Journal of Colloid and Interface Science* Vol. 361, No. 2 (September 2011) pp 618–627 ISSN 0021-9797

Part 3

Environmental Contamination and Genetics of Organisms (Uses and Prospects)

7

Bioindicator of Genotoxicity:
The *Allium cepa* Test

Solange Bosio Tedesco[1,2] and Haywood Dail Laughinghouse IV[3,4]
[1]Graduate Program in Agrobiology, Department of Biology
Universidade Federal de Santa Maria (UFSM), Santa Maria
[2]Graduate Program in Agronomy, UFSM, Santa Maria
[3]National Museum of Natural History, Smithsonian Institution, Washington, DC
[4]College of Computer, Mathematical & Natural Sciences
University of Maryland, College Park, MD
[1,2]RS Brazil
[3,4]USA

1. Introduction

In Brazil many species of medicinal plants are used in popular medicine. However, their indiscriminate and uncontrolled use can cause more harm to public health than good, thus the knowledge about these plants, from their cellular levels to their action on living organisms is important. The economic potential of the germplasm of medicinal species is an asset that should be preserved and developed, making it a more affordable alternative form of therapy for use by the general population, while conserving plant genetic diversity. To make this a reality, studies on the characterization of these plants on many levels, biological and/or agronomical, are essential, including the capacity of their extracts acting on other living organisms.

Cytogenetic studies of plant species report possible alterations of plant chromosomes due to mutagenic substances in their composition or resulting from their metabolism. The study of mutagens in eukaryotic nuclei has been observed by cytological methods and it is known that the mutation may result from the action of radiation, drugs and viruses, as well as the intrinsic stability of nucleic acids. Therefore, mutagens can be detected cytologically by cellular inhibition; disruption in metaphase; induction of chromosomal aberrations, numerical and structural, ranging from chromosomal fragmentation to the disorganization of the mitotic spindle, and consequently of all subsequent dependent mitotic phases.

Studies on the genotoxicity of medicinal plant extracts should be prioritized; in this way we invest research efforts towards public health. The analysis of chromosomal alterations serves as a mutagenicity test and is one of the few direct methods to measure damages in systems exposed to possible mutagens or carcinogens. To enable the evaluation of the effects or damages that mutagenic agents might cause, it is necessary for the sample to be in constant mitotic division, seeking to identify the toxic effects and alterations occurring over a cell cycle. In order to do so, there is the *Allium cepa* test, which has been widely used for this purpose (Silva et al., 2003).

The mitotic index and replication index are used as indicators of adequate cell proliferation (Gadano et al., 2002), which can be measured by the plant test system *Allium cepa*. Cytotoxicity tests, using plant test systems *in vivo*, such as *Allium cepa*, are validated by several researchers, who jointly performed animal testing *in vitro* and the results obtained are similar (Vicentini, et al. 2001; Teixeira et al. 2003), providing valuable information for human health. El-Shahaby et al. (2003) stress the importance of using the *Allium cepa* test for detecting toxicity/genotoxicity and evaluating environmental pollution.

As previously mentioned, in Brazil, medicinal plants are used for treating illnesses alternatively, and the test is important for alerting the general population about possible genotoxic risks, which can be caused in eukaryotic plant organisms, such as the onion. We intend to focus on the use of this test as a bioindicator of genotoxicity aiming at human health, demonstrating that even in an indirect way, it is possible to prevent and to avoid environmental contamination by the abusive use of substances that cause chromosomal aberrations. On the other hand, the results obtained through the application are surprising and show that certain plants are antimutagenic and would allow the reversion of genotoxic processes.

We emphasize the advantages of this test as a useful, low-cost system, as well as value the knowledge from plant cytogenetic techniques, which enable its development. Leme & Marin-Morales (2009) reported that studies of sensitivity and correlation of the *Allium cepa* (onion) test system and other test systems are fundamental for the more accurate evaluation of environmental risks, as well as the extrapolation of data to other organisms such as humans. The tests for risks to human health are performed by using various test systems and among these, the *Allium cepa* test.

2. Description and importance of the *Allium cepa* test

The *Allium cepa* test has been used by many researchers mainly as a bioindicator of environmental pollution (Bagatini et al. 2009; Leme & Marin-Morales, 2009), testing crude extracts of cyanobacteria (Laughinghouse, 2007), as well as to evaluate the genotoxic potential of medicinal plants (Camparoto et al. 2003; Knoll et al. 2006; Fachinetto et al. 2007; Lubini et al. 2008; Fachinetto et al. 2009; Fachinetto & Tedesco, 2009; Dalla Nora et al. 2010), because this test uses a model that is adequately sensitive to detect innumerous substances that cause chromosomal alterations.

The *Allium cepa* test is important since it is an excellent model *in vivo*, where the roots grow in direct contact with the substance of interest (i.e. effluent or complex medicinal mix being tested) enabling possible damage to the DNA of eukaryotes to be predicted. Therefore, the data can be extrapolated for all animal and plant biodiversity. The analysis of chromosomal alterations can be equal to the test of mutagenicity mainly for the detection of structural alterations; however, it is possible to observe numerical chromosomal alterations, as well. The *Allium cepa* test is one of the few direct methods for measuring damage in systems that are exposed to mutagens or potential carcinogens, and enables the evaluation of the effects of these damages through the observation of chromosomal alterations. For this undertaking, it is necessary that the sample remain in constant mitotic division, seeking to identify the toxic effects and alterations over a cell cycle; and the *Allium cepa* test has been widely used for this purpose. It is advantageous to use the *Allium cepa* test system since its main component is a vascular plant, making it an excellent genetic model for evaluating environmental pollutants, detecting mutagens in different environments and evaluating many genetic endpoints (point mutations to chromosomal alterations). *Allium cepa* is distinctive in regards to its efficiency in detecting genetic damage and was introduced by Levan, in 1938, for helping observe disturbances in the mitotic fuse due to colchicine action.

Relevant studies by Fiskesjö (1985), showed the importance of the *Allium cepa* test system for evaluating genotoxicity, demonstrating that *Allium cepa* cells contain an oxidase enzyme system capable of metabolizing polycyclic hydrocarbonates. Even though other test systems have been shown to be sensitive for this detection, the results of the *Allium cepa* test should be considered as an alert for other organisms (i.e. bioindicators). Studies on sensibility and correlation among test systems are fundamental for a more accurate evaluation of environmental risks and for extrapolating the data to other groups of target organisms. A high sensitivity and good correlation with mammal tests and the same sensitivity as test systems of algae and human lymphocytes exist when compared with *Allium cepa*.

Furthermore, Rank & Nielsen (1993) performed adaptations for evaluating complex mixtures and Ma et al. (1995) adjusted the test for assessing mutagenicity and micronucleus analyses (MN) in F1 cells. Some researchers show certain restriction in regards to using plant test systems for evaluating certain classes of carcinogens, which require complex metabolization systems for the activation of its genotoxic action. However, Rank & Nielsen (1994) showed a correlation of 82% between the *A. cepa* test and the carcinogenicity test in rodents and concluded that the same was even more sensitive than the Ames test. Vincentini et al. (2001) reported that the *Allium cepa* test system is well accepted for the analysis of cytotoxicity and genotoxicity because the roots are in direct contact with the tested substance, allowing evaluation of different concentrations and times. The results by Camparoto et al. (2002) were similar when estimating the effects of infusions of *Maytenus ilicifolia* Mart. and *Bauhinia candicans* Benth with the test of *Allium cepa* and bone marrow cells of Wistar rats. Studies by Knoll et al. (2006) used the plant model of *Allium cepa* to test several populations of *Pterocaulon polystachyum* DC (known as quitoco in the southern region of Brazil) at different concentrations and obtained precise results on cell division inhibition, whose effects were attributed to the presence of flavonoids in the infusions tested and the authors demonstrated that with an increase in the concentration of the infusions of *P. polystachyum*, lower mitotic index values were recorded.

The use of medicinal plants for treating illnesses is an exploratory practice that is widely diffused in Brazil (Rosa & Ferreira, 2011) and due to this intense medicinal use, studies using bioindicators of toxicity and mutagenicity, such as the *in vivo* test of *Allium cepa* are necessary for contributing to their safe and efficient use. The plant test system of *Allium cepa* is as an ideal bioindicator for the first *screening* of genotoxicity, helping with studies that prevent damages to human health (Bagatini et al., 2007). Lubini et al. (2008) studied two species of the genus *Psychotria, Psychotria leiocarpa* Cham. & Schltdl. and *Psychotria myriantha* Mull. Arg. Using *Allium cepa* to test infusions at two concentrations of these species it was possible to verify the antiproliferative activity of the species *P. leiocarpa* and *P. myriantha*, and the results indicated that both species possessed the capacity to inhibit cell division and *P. myriantha* possessed genotoxic activity. We can indicate the use of *P. leiocarpa* in high concentrations as potentially therapeutic for inhibiting the cell cycle in eukaryotic organisms.

Studies by Souza et al. (2010) demonstrated that the species *Artemisia verlotorum* Lamotte (known in Brazil as infalivina) has antiproliferative and genotoxic capacity on the *Allium cepa* cell cycle. The authors found 32.2% of the chromosomal alterations in the highest concentration of the tested aqueous extract, 48 g/L. It was found that with an increase in the aqueous extract of *A. verlotorum*, there was higher inhibition of cell division, and consequently lower values of MI.

3. Methodology of the *Allium cepa* test

The *Allium cepa* test consists in obtaining onion bulbs cultivated without the application of herbicides or fungicides. After obtaining the bulbs, they should be scraped at the root to promote the emergence of new roots. Other bioassays can be performed with *Allium cepa* seeds placed to germinate in a BOD (biochemical oxygen demand) incubator with controlled temperature, which are used for allelopathy testing and also genotoxicity assessment. To set-up the experiment allowing rootlets to grow, all bulbs should be placed initially in a small 50 mL plastic cup (Figure 1), containing distilled or tap water (being that it is potable), for approximately 03 to 04 days so rootlets can emerge. After this period, the bulbs should be transferred to other clean and dry containers including the treatments. The plastic cups for water control treatment can be reused, to minimize the effect of plastic on the environment. In general, we use 5 groups of bulbs of *Allium cepa* for each treatment, with one being a negative control in water and another for the positive control in methylsulfonylmethane (MMS) or glyphosate. The issue of positive control is cited by Rank (2003) using methylsulfonylmethane and by Souza et al. (2010) using glyphosate. Through the studies developed to date at the Laboratory of Plant Cytogenetics and Genotoxicity (LABCITOGEN) of the Department of Biology at the Federal University of Santa Maria, various concentrations of glyphosate have shown chromosomal changes directly to the rootlets. Its residual effect on the environment after us is not proven. After rooting, the bulbs of the two control groups should remain in the water (negative control) and in the respective positive control, and the remaining should be transferred to the chosen treatments, which can be solutions of essential oil, leaf extracts by infusion, root or stem extracts by decoction, or samples of industrial and/or hospital effluents. These should remain in the dark for exactly 24 hours.

It is emphasized that in a case where a product such as ethanol PA is used for the dilution of the oil in which the rootlets of *Allium cepa* will be immersed, the same should consist of one of the treatments. The main idea is to minimize the error, allowing the results to express the effect of the substances that are interacting. Therefore, when possible, test them separately. After the rootlets are submitted for 24 hours in the individual treatments, they should be collected and immediately fixed in ethanol: acetic acid (3:1), also for 24 hours.

Afterwards, the rootlets should be removed from the fixing solution and transferred to ethanol 70% and be kept under refrigeration (4°C) until used. It is important to emphasize that all glassware used to keep the rootlets should be identified with a number or sample name and/or treatment, as well as date, using small tags written in pencil on the inner and outer part of the glass, thus avoiding any kind of mistake with the samples. The technique of bulb use is described by Fiskesjö (1994) and adapted by researchers who use the test, such as Knoll et al. (2006) and Fachinetto et al. (2009). In the next stage, slides are prepared, analyzing 1000 cells per bulb, totaling 5000 cells per treatment or variations of these values, such as 500 cells per bulb, totaling 2500 cells. It has already been proven that one rootlet is enough for observing the damage caused to the DNA of *Allium cepa*, observing the cells after the treatment with mutagenic agents.

The researcher should have a large sample so that errors can be minimized. We believe that at least 200 cells be analyzed per bulb, totaling 1000 cells per treatment, in case of optimizing time, considering such samples as pilot or initial experiment. Slide preparation can follow the technique by Guerra & Souza (2002) where the whole technique of squashing and staining of the root tip for obtaining cells with good visualization is explained. In the routine of LABCITOGEN, for slide preparation, the rootlets are hydrolyzed in HCl for 5-6 minutes and rinsed in distilled water. Then, only the meristematic region of the rootlets (eliminating the rest and the root cap) should be fragmented, under the lens of a stereoscopic microscope using histological needles, previously stained with 2% acetic orcein. For Raven et al. (1988) the distance in relation to the apical meristem, where we find the location with the majority of cell division varies among species, and within a single species, depending on the root age. The combination of apical meristem plus the portion of the root where the cell divisions occur is called the cell division region or zone of division, including the quiescent region. In relation to staining, other stains can also be used, such as acetic orcein 1%, acetic carmine 1 and 2%, as well as Giemsa. During squashing and staining, a cover slip is used and carefully squeezed with filter paper to remove excess stain. Then, holding one end of the cover slip, it should be tapped with the end of the histological needle several times to spread the cells. Excessive force when tapping can tear the cells; this should not be confused with a morphological alteration.

Fig. 1. Sample of an onion bulb showing the rootlets already grown. Arrow indicates the plastic cup, in which the bulb can be placed for testing. Scale bar=2cm

Fig. 2. Organization of the experiment with *Allium cepa*, containing 4 groups of 5 onion bulbs, LABCITOGEN, UFSM, Santa Maria, RS, Brazil. Scale bar=2cm

The slides are then evaluated using a LEICA microscope with 400X magnification and the cells are observed during the interphase and cell division, in prophase, metaphase, anaphase, and telophase. Cell counts are carried out considering visual fields scanning the whole slide. The counts should be written down on a table, which is shown in Table 1. Following, the sum of all cells in interphase and in division is carried out and then the mitotic index (MI) is calculated for each treatment. During the cell count, they should be divided into two categories: regular (do not present damages in the chromosomes) and irregular (present damages in the chromosome, such as: chromosomal breakages, simple or multiple anaphasic bridges, micronucleus, laggard or lost chromosomes). The statistical analysis should be preferably performed using the χ^2 test with a $p<0.05$ level of probability using a statistical program, e.g. BioEstat 4.0 (Ayres & Ayres, 2003) or BioEstat 5.0 (Ayres & Ayres, 2009). The mitotic index can be calculated through the number of cells observed in prophase + metaphase + anaphase + telophase (Love & Love, 1975). According to Sehgal et al. (2006), the mitotic index can be recorded using the formula:

$$\text{Mitotic index} = \frac{P + M + A + T}{\text{Total number of cells}}$$

For Ozmen & Summer (2004) the mitotic index was expressed in terms of divided cells/total number of cells and in this study, it was observed that the chromosomal aberrations were determined by scoring cells with bridges, fragments, sticky chromosomes and polar deviation in three randomly picked zones and micronucleus formation in 1,000 cells per slide. In their study, Ozmen & Summer (2004) used *Allium cepa* to test extracts of *Plantago lanceolata*. The researchers used two groups of ten bulbs which were placed to germinate in the dark at 22°C, and the same were observed after 48h, removing the bulbs which had roots that only developed a little, and the rest were treated with water with 0.7% H_2O_2 for 1h. At the end of the treatment with H_2O_2, the roots were washed and then treated with two different concentrations of extracts (15 g/L and 30 g/L) of *Plantago lanceolata* for 24 hours. In this study the onions were placed to root in the dark, because the plant extracts were photosensitive. In the same way, other researchers placed the onions to root in the dark (Fachinetto et al., 2008) to test nanocoated tretinoin, since this was also photosensitive. So, the *Allium cepa* test can be undertaken with rooting the bulbs in either the dark or light.

Rank & Nielsen (2003) explain in their study the anaphase-telophase method of *Allium cepa* and made important considerations, such as genotoxic chemicals used for many purposes in manufacturing processes that can be found in environmental compartments such as air, water, soil, and sediments. The chemical can enter the environment from discharged wastewater, air emissions, during product consumption and from domestic and industrial waste sites. The important advantage of the *Allium cepa* test is that it is a "low budget" method, which besides being fast and easy to handle also gives reliable results.

To analyze cell division in *Allium cepa*, it is necessary to correctly identify all the phases that are considered for the calculation of the mitotic index. In Figure 3, there are cells in interphase and in cell division. Normally, most cells are found in interphase, thus the mitotic index value obtained from the negative control in water is variable.

The regular phases of *Allium cepa* cell division are presented and shown in Figure 4-A, B, C and D. In these phases of mitosis, there is an example of regular cells when the roots only grow in distilled water, not presenting chromosomal abnormalities. Figure 4-A shows prophase with chromosomes quite visible, which is characteristic of this phase; 4-B is metaphase with chromosomes arranged in the equatorial plate of the cell awaiting subsequent movement to the opposite poles during anaphase; 4-C shows anaphase with the chromosomes moving to the opposite poles of the cell in a stable way; 4-D is the phase that finalizes the mitotic cell division, telophase, showing the chromosomes already organized on opposite poles of the cell, awaiting the following step of cytokinesis, forming 2 new daughter cells.

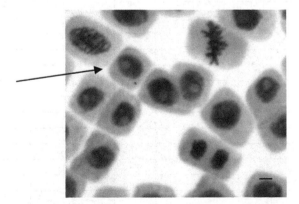

Fig. 3. *Allium cepa* cells obtained from 2cm-rootlets, after 24h under control in distilled water, fixed, stained and prepared according to the routine of LABCITOGEN, UFSM, Santa Maria, RS, Brazil. The arrow indicates cells in interphase. Scale bar= 10μ

Slide	Number of normal or regular cells observed					Number of abnormal or cells with chromosomal irregularities				
	Inter	Pro	Meta	Ana	Telo	Inter	Pro	Meta	Ana	Telo
S1B1T1										
S2B1T1										
S3B2T1										
S4B2T1										
S5B3T1										
S6B3T1										
S7B4T1										
S8B4T1										
S9B5T1										
S10B5T1										
S1B1T2										
S2B1T2										
S3B2T2										
S4B2T2										
S5B3T2										
S6B3T2										
S7B4T2										
S8B4T2										
S9B5T2										
S10B5T2										
S1B1T3										
S2B1T3										
S3B2T3										
S4B2T3										
S5B3T3										
S6B3T3										
S7B4T3										
S8B4T3										
S9B5T3										
S10B5T3										
S1B1T4										
S2B1T4										
S3B2T4										
S4B2T4										
S5B3T4										
S6B3T4										
S7B4T4										
S8B4T4										
S9B5T4										
S10B5T4										

S1B1T5											
S2B1T5											
S3B2T5											
S4B2T5											
S5B3T5											
S6B3T5											
S7B4T5											
S8B4T5											
S9B5T5											
S10B5T5											

S=Slide, B=bulb, T=treatment, Inter=interphase, Pro=prophase, Meta=metaphase, Ana=anaphase, Telo=telophase

Table 1. Illustrative Framework for use in annotating results of an experimental design with 5 bulbs of *Allium cepa* and 5 distinct treatments

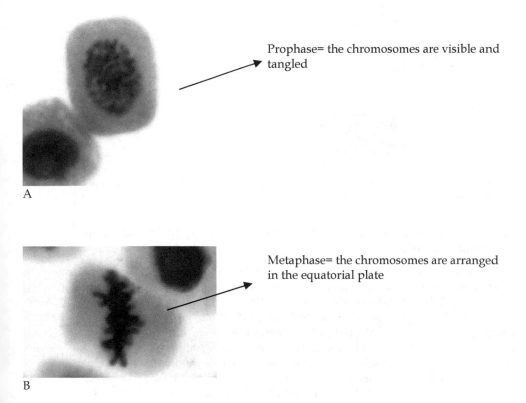

Prophase= the chromosomes are visible and tangled

A

Metaphase= the chromosomes are arranged in the equatorial plate

B

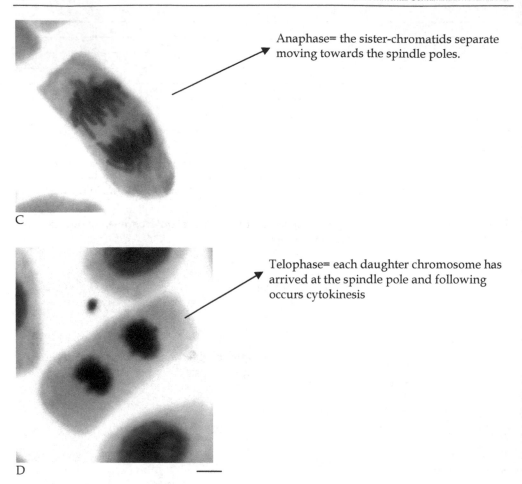

Anaphase= the sister-chromatids separate moving towards the spindle poles.

Telophase= each daughter chromosome has arrived at the spindle pole and following occurs cytokinesis

C

D

Fig. 4. *Allium cepa* cells in regular or normal division. A-interphase, B- prophase, C-metaphase, D- telophase. Scale bar= 10μ

4. Use as a bioindicator in detecting the genotoxicity of hospital and industrial effluents

Cytogenetic tests are desirable for identifying the damaging effects of substances known in various concentrations under different exposure times for evaluation and influence on living organisms (Al-Sabti & Kurelec, 1985; A1-Sabti, 1989; Abdou et al., 1989; Arrigoni et al., 1989; Kakand Kaul, 1989; Kumar & Sinha, 1989; Rao, 1989; Chauhan & Sunderaraman, 1990; EI-Khodary et al., 1990; Panda et al., 1990; Singh et al., 1990; Kumar et al., 1991; De-Serres, 1992). These tests provide data for understanding the harmful effects on tested organisms and are commonly used for biomonitoring pollutants, in addition to evaluating the effects of toxic and mutagenic substances on organisms in their natural habitat (Degrassi & Rizzoni, 1982; Al-Sabti & Kurelec, 1985; Dixit & Nerle, 1985; Fiskesjö, 1988; De Marco et al., 1988; Al-Sabti, 1989).

Industrial effluents have become one of the biggest problems in many developing and developed countries. It is known that these effluents, when not treated properly, can cause mutagenic or toxic effects directly on humans, affecting human health, resulting in diseases, such as cancer, congenital malformations, and cardiovascular diseases (Grover & Kauer, 1999). Studies by Siddiqui et al. (2011) were undertaken to validate plant-based tests for assessing the toxicity of water in India. In this study, the authors reported that plant-based bioassays have become increasingly popular for toxicological and eco-toxicological evaluations, and that the main reasons for the widespread use of the methods are simplicity, sensitivity and low cost, as well as the positive correlation with other toxicity tests.

Leme & Marin-Morales (2009) carried out an extensive review on the *Allium cepa* test and its use in environmental contamination, where they reported that vascular plants are recognized as excellent genetic models for detecting environmental mutagens and are frequently used in monitoring studies. *Allium cepa* is among the plant species used to evaluate DNA damages, chromosomal alterations and disturbances in the mitotic cycle. Furthermore, they reported that the test has been used to evaluate a large number of chemical agents, increasing its environmental application and it is a test characterized by being cheap. They also commented how the *Allium cepa* has advantages over other tests by the short preparation time for testing samples and although plants have low concentrations of oxidase enzymes, their results are consistent and can serve as a warning for other biological systems, since the target is DNA, which is common in all. In this review, they demonstrate that all types of effluents are also considered as complex mixtures and that the main results are of cytotoxicity, genotoxicity, and mutagenicity. In their review, Leme & Marin-Morales (2009) summarized data from several researchers on a large range of environmental contaminants and their genotoxic, cytotoxic or mutagenic effects on *Allium cepa*, such as pesticides, herbicides, metals and heavy metals.

Hospital effluents can cause severe problems to live organisms when not properly treated, and in developing countries, such as Brazil, environmental contamination by these effluents is not uncommon. This contamination is due to mutagenic compounds found within the effluent. Biomonitors (i.e. *Allium cepa*) can be used to alert the surrounding population of environmental contamination and genotoxic substances that have been released into the water. In a study by Bagatini et al. (2009), the *Allium cepa* test was used to evaluate the genotoxicity of a hospital effluent in Santa Maria, Rio Grande do Sul State, Brazil. During the study, chromosomal disruptions, anaphasic bridges, and micronuclei during telophase were observed, indicating environmental toxicity risk.

Laughinghouse (2007) studied the cytotoxic effects of crude extracts of cyanobacteria (blue-green algae), which can cause water pollution and the damages of direct toxin-producing strains can be tested using the *Allium cepa* test. The comparison of the toxic and genotoxic effects among species is fundamental for evaluating the biological risk of pollutants, particularly for compounds persistent in the environment (Bolognesi et al., 1999). The occurrence of blooms in continental waters used for human consumption causes essentially two problems for treatment. On one hand, being very small organisms, they can pass through filters at Water Treatment Plants (WTPs), reaching high densities in the distribution network. On the other, their toxins are not removed by the usual treatments (coagulation, flocculation, filtration, and disinfection), being even resistant to boiling. Besides these aspects, the traditional treatments can increase the risk of forming organochlorine compounds from the group of trihalomethanes, which act as carcinogenic compounds when water rich in organic matter is treated with chlorine. It is important when there is a higher density of toxic cyanobacteria, not to resort to using pre-chlorination, but to use activated charcoal filters and ozone, which remove the toxins in the water more efficiently (Schmidt et al., 2002; Antoniou et al., 2005; Azevedo, 2006). The accidental ingestion of waters with high levels of toxins (acute ingestion) can cause intoxications, characterized by gastroenteritis with diarrhea, vomiting, nausea, abdominal cramps and fever, hepatitis with anorexia, asthenia and vomiting, or death (Jochimsen et al., 1998). The continued ingestion of low doses of toxins (chronic ingestion) can lead to chronic liver disorders. In fact, there should be me more studies showing what are the risks of chronic exposure are since there are many factors that can lead a person to be subjected to low doses of these toxins. These situations can also be triggered by the ingestion of mollusks, as filter feeders that accumulate non-lethal doses of these products in their tissues, which are passed along the food chain, finally to humans (Kuiper-Goodman et al., 1999; Azevedo, 2006; Carvalho, 2006).

The decrease in water quality, especially of environments used for public water supply, irrigation and recreation, is of concern. The increase in eutrophication in these systems by higher nutrient loads (especially phosphorus and nitrogen) have favored the predominance of toxigenic cyanobacteria threatening human and animal health, aside from elevating the cost of water treatment. Thus, as a result of eutrophication, many countries, have suffered from an increase of toxic blooms, which is a severe problem to public health (Werner & Laughinghouse IV, 2006).

5. Use as a 'warning' bioindicator in detecting genotoxicity of medicinal plants

Various species of medicinal plants are used in popular medicine for the treatment of illnesses. However, the presence of cytotoxic and mutagenic substances in their composition or resulting from their metabolism can cause damage to human health. The mutagenic effects result in chromosomal alterations detected during the cell cycle through cytogenetic analysis. What does genotoxicity mean? It refers to the capacity of clastogenic agents causing lesions in the genetic material. Genotoxic agents can be defined functionally for possessing the ability to alter DNA replication and genetic transmission. The evaluations of genotoxicity include, mainly, damage in the DNA, mutations and chromosomal alterations. The observation of cells in interphase and cell division is used as an indicator of adequate proliferation of the cells, which can be measured through the *Allium cepa* test system.

The studies by various authors, such as Vicentini et al. (2001) and Camparoto et al. (2002) were performed with the *Allium cepa* test to test the genotoxicity of complex mixtures, in reality known as teas or extracts. The evaluation of the genotoxic effects of the plant extracts has been studied by Chauan et al. (1999), who indicated the sensitivity of the *A. cepa* system and correlated it with the mammal test system, validating its use as an alternative test for monitoring the potential genotoxicity of environmental chemicals and pesticides.

Meristematic onion cells and rat cells were used as test systems to verify the effects of genotoxicity of extracts (infusions) of medicinal plants such as *Maytenus ilicifolia* Mart and

Bauhinia candicans Benth. by Camparoto et al. (2002) demonstrating that there was not a significant difference in the decrease of the mitotic index in both cases studied; there was only a decrease in the mitotic index of the meristematic cells in the onion, whose bulbs were treated with a higher concentration (10 x higher than the one used by the population in the form of medicinal tea) of *Bauhinia candicans*. These studies indicated that the use of these plants could be continued, as long as they are always used in the recommended dosage.

Lubini et al. (2008) analyzed the genotoxicity of two species of *Psychotria* (*P. leiocarpa* and *P. myriantha*) through the *Allium cepa* test and the results indicated that both species possess capacity to inhibit cell division and *P. myriantha* possesses genotoxic activity. Çelik et al. (2006) studied extracts of *Plantago lanceolata* L. and their results showed that aqueous extracts reduced mitotic index and chromosome aberrations in treatment groups compared to controls. The results of the presented study are therefore important since they suggested the anti-genotoxic effect of the *P. lanceolata* leaf extract. In order to reach certain conclusions about this subject, however, further research should be performed with different test systems.

Extracts are most commonly prepared by infusion or decoction, depending on the part of the plant used. In infusion, extraction is carried out when the plant material is maintained in boiling water, in a covered container, for a certain period of time. Infusions can be applied to plant parts of soft structure, which should be beaten, cut or pulverized roughly according to their nature so that they can be easily penetrated and extracted with water. However, decoction consists of maintaining the plant material in contact, during a certain period of time, with a boiling solvent (usually water). It is a technique of restricted use, since many active substances are altered by a prolonged period of heating and it is customary to employ it with rigid/woody plants (Simões et al., 2011).

Worldwide, many species of medicinal plants are used to treat illnesses. Most of these species are not thoroughly studied, especially regarding the presence of toxic/mutagenic substances in their composition or arising from their own metabolism, thus damaging the health of the population. The presence of mutagenic substances in the plant species that might cause chromosomal alterations can be detected during the cell cycle of a species. The *Allium cepa* test system is frequently used for evaluating the potential genotoxicity of medicinal plant extracts through the analysis of meristematic cells from root-tips treated with medicinal infusions (teas). The knowledge of the potential genotoxicity of these medicinal species, through the analysis of the *Allium cepa* cell cycle serves as an indicator of safety for the population, which uses medicinal teas as their only medical treatment. Bagatini et al. (2007) reviewed this subject and indicated the importance of the *Allium cepa* test as a preliminary screening of genotoxicity in medicinal plant infusions.

6. Types of results and interpretation through the analysis of plant cytogenetics

The results, which can be obtained by analyzing the rootlets subjected to different treatments of interest to researchers, are performed by cytogenetic analysis during cell division. The *Allium cepa* cell cycle can be taken into consideration after 24hrs, and is divided into interphase and cell division, understanding the prophase, metaphase, anaphase, and telophase phases. Some authors, such as Singh (2002) consider all phases of cell division starting with interphase. In the case of the *Allium cepa* test for the interpretation of the results, the subdivision is necessary, because those cells which are in interphase are not considered as cells in division. According to Love (1949) the mitotic index can be calculated from the formula:

$$MI = \frac{\text{total number of cells observed (cells in interphase + number of cells in division)}}{\text{number of cells in interphase}} \times 100$$

To get a percentage, multiply the result by 100. The slide must be prepared from the meristematic region (by removing the root cap) and the squashing technique should be used (Guerra & Souza, 2002), soon after stained with acetic orcein 2% or another stain that has the same affinity for DNA packaged as chromosomes. The observation of the cells through the microscope can be interpreted based on the regular cell division of *Allium cepa*. Figure 5-A, 5-B, 5-C, 5-D, 5-E, 5-F represents the phases involved in the analysis of irregular mitotic division, being respectively, A and B- irregular anaphase, C- irregular metaphase , D-binucleate cell, E-adherent cell, and F- binucleate cell.

Depending on the tested substances, such as, herbicides used in agricultural practice or leftover drugs discarded after use, it is possible to observe if these substances are mutagenic or even antimutagenic. If they are mutagenic it is possible to immediately see this through the structural damages such as those in Figures 6-A, 6-B, 6-C, 6-D, 6-E, 6-F, 6-G, where there are chromosomes with breaks, simple anaphasic bridges, multiple anaphasic bridge, adhesions, laggard chromosomes, disorganization of the metaphase, and binucleated cells. If they are antimutagenic it is necessary to verify the reversion of the mutations occurred.

The use of antimutagens and anticarcinogens every day is the most efficient procedure to prevent human cancer and genetic illnesses. There are many ways in which the action of mutagens can be reduced or prevented. Ragunathan & Paneersel Vam (2007) studied the antimutagenic potential of curcumin on chromosome aberrations in *Allium cepa*. These authors recorded that turmeric has long been used as a spice and food coloring agent in Asia. The antimutagenic potential of curcumin was evaluated using *Allium cepa* root meristem cells. The authors found that curcumin insignificantly induced chromosomal aberrations. However, the authors noticed that this spice had an antimutagenic potential against sodium azide (known to induce chromosomal aberrations). Thus, the mechanism of action remains unknown, though curcumin presented an antimutagenic potential demonstrated by the *Allium cepa* test.

Other important studies, such as Rossato et al. (2010), took into account the different cultivation environments of medicinal plants used for alternative medicine and then tested by *Allium cepa*. These authors evaluated the antiproliferative effect of *Pluchea sagittalis* (Lam.) Cabrera infusions at three different concentrations (2.5, 5 and 25 gdm-³) of leaves of *P. sagittalis* grown in three environments (in vitro, greenhouse, and field). From this study, they concluded that the leaf infusions of the studied species have a large proliferative capacity and the plant cultivation system affects the values of mitotic index, i.e. the proliferative capacity of cells.

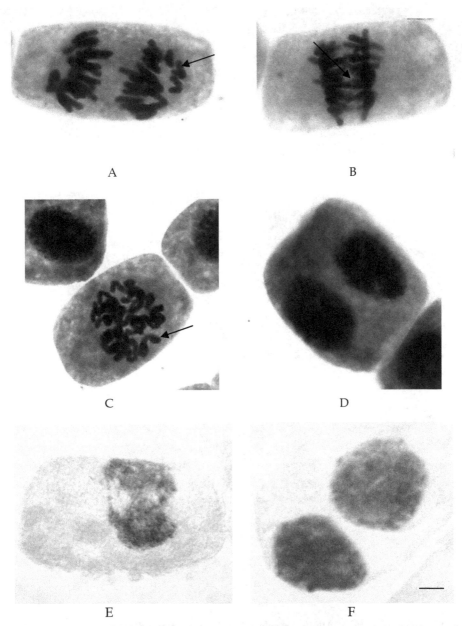

Fig. 5. *Allium cepa* meristematic cells showing the alterations due to the action of clastogenic agents. A- irregular anaphase, Arrow indicating lost chromosomes, B- irregular anaphase, with anaphasic microbridges, C- irregular metaphase, with unorganized chromosome, also known as C-metaphase, showing chromosomes with no orientation on the equatorial plate, D- irregular cell, binucleate, with an elliptical aspect, E- adherent nucleus, F- irregular cell, binucleate, with a round aspect. Scale bar= 10µ

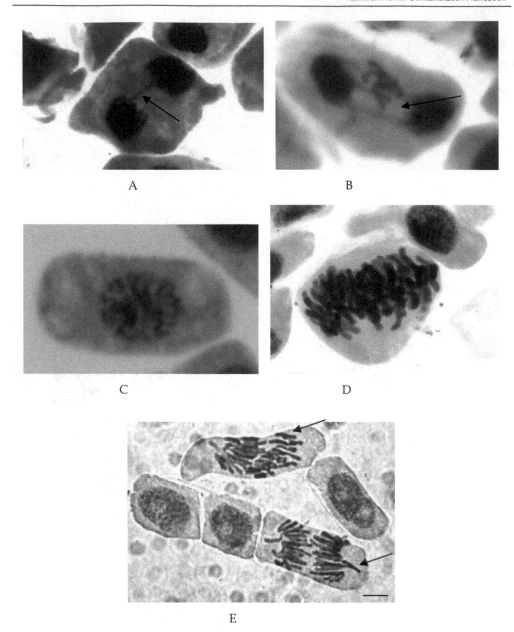

Fig. 6. Meristematic cells of *Allium cepa* showing the alterations due to clastogenic agents. A- cell in irregular anaphase, arrow indicates anaphasic bridge, B- anaphase with a bridge and arrow indicating lost chromosomes, C- irregular prophase, showing decompressed chromosomes, D- metaphase with numerical alteration, due to duplication of the number of chromosomes, E- cells in irregular anaphase, arrows indicating chromosomal breakage. Scale bar=10μ

7. Conclusions

We conclude that the *Allium cepa* test is an excellent bioindicator of chromosomal alterations that serve as an alert for the population that uses medicinal teas indiscriminately, and that its constant use in the analysis of the treatment of industrial and hospital effluents is extremely adequate. Currently, due to major concern with environmental pollution, the *Allium cepa* test has occupied an important place for the prevention and prediction of environmental impact that will be caused by the use and disposal of substances including drugs and herbicides.

Although the test is merely a first assessment of genotoxicity, it always shows important scientific discoveries, and new adaptations of the test might reveal innumerous possibilities of its use, avoiding the use of animals for testing. More increments and analysis, as the sophistication of the method progresses, will lead us to get the most use for the benefit of the planet.

8. Acknowledgements

We would like to extend our thanks to all the students that have passed through LABCITOGEN, and helped with analyzing *Allium cepa*. We are also indebted to Richard M. Fischer for support in reviewing the final text.

9. References

Abdou, R.F., S.E. Megalla, A.M. Moharram, K.M. Abdal, T.H.I. Sherif, A.L. El-Syed-Mahmood & A.E. Lottfy. (1989). Cytological effects of fungal metabolites produced by fungi isolated from Egyptian poultry feedstuffs, *J. Basic. Microbiol.*, 29, pp. 131-139, ISSN 1521-4028

Al-Sabti, K. & Kurelec, B., (1985). Induction of chromosomal aberrations in the mussel *Mytilus galloprovincialis* watch, *Bull. Environ. Contam. Toxicol.*, 35, pp. 660–665, ISSN 1432-0800

Antoniou, M.G., Cruz, A.A. de la & Dionysiou, D.D. (2005) Cyanotoxins: New Generation of Water Contaminants, *Journal of Environmental Engineering*, pp. 1239-1243, ISSN 0733-9372.

Ayres, M. & Ayres Jr, M. (2003). *BioEstat 3.0: aplicações estatísticas nas áreas das ciências biológicas e médicas,* Sociedade Civil Mamirauá, Belém, PA.

Ayres M., Ayres Jr, M., Ayres D.L. & Santos, A.S. (2005). *Bioestat 4.0: aplicações estatísticas nas áreas das ciências biomédicas,* Sociedade Civil Mamirauá, Belém, PA.

Azevedo, M.T. de P. & Sant'Anna, C.L. (2006). Morfologia e reprodução, In: *Manual Ilustrado para Identificação e Contagem de Cianobactérias Planctônicas de Águas Continentais Brasileiras,* Sant'Anna, C.L., Azevedo, M.T. de P., Agujaro, L.F., Carvalho, M. do C., Carvalho, L.R. & Souza R.C.R. de, pp. 5-8, Interciência, Rio de Janeiro.

Bagatini, M.D., Fachinetto, J.M., Silva, A.C.F. & Tedesco, S.B. (2009). Cytotoxic effects of infusions (tea) of *Solidago microglossa* DC. (Asteraceae) on the cell cycle of *Allium cepa*, *Brazilian Journal of Pharmacognosy*, 19(2B), pp. 632-636, ISSN: 0102695X

Bagatini M.D., Silva A.C.F. & Tedesco S.B. (2007).   Uso do sistema teste de *Allium cepa* como bioindicador de genotoxicidade de infusões de plantas medicinais. *Brazilian Journal of Pharmacognosy*, 17, pp.444-447, ISSN 0102695X

Bagatini, M.D., Vasconcelos, T.G., Laughinghouse IV, H.D. Martins, A.F. & Tedesco, S.B. (2009).   Biomonitoring Hospital Effluents By *Allium cepa* L. Test. *Bulletin of Environmental Toxicology and Contamination*, 82, pp. 590-592, ISSN 0007-4861

Bolognesi, C., Landini, E., Roggieri, P., Fabbri, R. & Viarengo, A. (1999).   Genotoxicity biomarkers in the assessment of heavy metal effects in mussels: Experimental studies. *Environmental Molecular Mutagenesis*, 33, pp. 287-292, *ISSN* 0893-6692.

Cabrera, G.L. & Rodriguez, D.M.G. (1999).   Genotoxicity of soil from farmland irrigated with wastewater using three plant bioassays, *Mutat. Res.*, 426, pp. 211–214, ISSN 0027-5107

Camparoto, M. L., Teixeira, R.O., Mantovani, M. S. & Vicentini, V.E.P. (2002).   Effects of *Maytenus ilicifolia* Mart. and *Bauhinia candicans* Benth infusions on onion root-tip and rat bone-marrow cells. *Genetics and Molecular Biology*, 25, pp. 85-89, ISSN 1415-4757

Carvalho, L.R. (2006).   Cianotoxinas, In: *Manual Ilustrado para Identificação e Contagem de Cianobactérias Planctônicas de Águas Continentais Brasileiras*, Sant'Anna, C.L., Azevedo, M.T. de P., Agujaro, L.F., Carvalho, M. do C., Carvalho, L.R. & Souza R.C.R. de, pp. 9-19, Interciência, Rio de Janeiro.

Chauhan, L.K.S., Saxena P.N. & Gupta, S.K. (1999).   Cytogenetic effects of cypermethrin and fenvalerate on the root meristem cells of *Allium cepa*, *Environmental and Experimental Botany*, 42, pp. 181-189, ISSN 0098-8472

Çelik, T.A. & Aslantürk, O.S. (2006). Anti-mitotic and anti-genotoxic effects of *Plantago lanceolata* aqueous extract on *Allium cepa* root tip meristem cells, *Biologia*, 61, 6, pp. 693 – 697, ISSN 1336-9563.

Degrassi F. & Rizzoni M. (1982).   Micronucleus test in *Vicia fabra* root tips to detect mutagen damage in fresh-water pollution, *Mutation Research*, 97, pp. 19-33, *ISSN*: 0027-5107.

Dixit, G.B. & Nerle, S.K. (1985).   Cytotoxic effects of industrial effluents on *Allium cepa* L., *Geobios*, 12, pp. 237-240, *ISSN* 0251-1223

El-Khodary, S., Habib, A. & Haliem, A. (1990).   Effect of the herbicide tribunil on root mitosis of *Allium cepa*, *Cytologia*, 55, pp. 209-215, ISSN 0011-4545

El-Shahaby A.O., Abdel Migid H.M., Soliman M.I. & Mashaly I.A. (2003).   Genotoxicity screening of industrial wastewater using the *Allium cepa* chromosome aberration assay, *Pakistan Journal of Biological Sciences*, 6, 1, pp. 23-28, ISSN 1812-5735

Fachinetto, J.M., Bagatini, M.D., Silva, A.C.F. & Tedesco, S.B. (2007).   Efeito anti-proliferativo das infusões de *Achyrocline satureioides* DC (Asteraceae) sobre o ciclo celular de *Allium cepa*, *Rev. bras. farmacogn.*, 17, 1, pp. 49-54, ISSN: 0102695X

Fachinetto, J.M., Ourique, A., Lubini, G., Tedesco, S.B., Silva, A.C.F. & Beck, R.C.R. (2008).   Tretinoin-loaded polymeric nanocapsules: evaluation of the potential to improve the antiproliferative activities on *Allium cepa* root-tip compared to the free drug, *Acta Farmaceutica Bonaerense*, 27, pp. 668-673

Fachinetto, J.M. & Tedesco, S.B. (2009).   Atividade antiproliferativa e mutagênica dos extratos aquosos de *Baccharis trimera* (Less.) A. P. de Candolle e *Baccharis articulata* (Lam.) Pers. (Asteraceae) sobre o sistema teste de *Allium cepa*, *Rev. Bras. Pl. Med.*, 11, 4, pp. 360-367, ISSN 1516-0572

Fiskesjö, G. (1985). The *Allium*-test as a standard in environmental monitoring, *Hereditas*, 102, pp. 99-112.

Fiskesjö, G. (1988). The *Allium* test-an alternative in environmental studies: The relative toxicity of metal ions. *Mutation Research*, 197, pp. 243–260, ISSN 0027-5107

Fiskesjö, G. (1993). The *Allium* test in wastewater monitoring, *Environ Toxicol Water Qual*, 8, pp. 291-298.

Fiskesjö, G. (1994). The *Allium* Test II: Assesment of chemical's genotoxic potential by recording aberrations in chromosomes and cell divisions in root tips of *Allium cepa* L., *Environ Toxicol Water Qual*, 9, pp. 234-241.

Fiskesjö, G. (1995). *Allium* test: In vitro toxicity testing protocols, *Meth. Mol. Biol.*, 43, pp. 119-127.

Gadano, A., Gurni, A., López, P., Ferraro, G. & Carballo, M. (2002). In vitro genotoxic evaluation of the medicinal plant *Chenopodium ambrosioides* L., *J Ethonopharmacol*, 81, pp. 11-16.

Grover, I.S. & Kaur, S. (1999) Genotoxicity of wastewater samples from sewage and industrial effluent detected by the *Allium* root anaphase aberration and micronucleus assays, *Mutat. Res.*, 426, pp. 183–188, ISSN 0027-5107

Guerra, M. & Lopes, M.J.S. (2002). *Como observar cromossomos - Um guia de técnicas em citogenética vegetal, animal e humana* (1). Ed. FUNPEC, Ribeirão Preto, SP.

Hunter, P.R. (1998). Cyanobacterial toxins and human health. *Journal of Applied Microbiology*, 84, pp. 35S-40S.

Jochimsen, E.M., Carmichael, W.W., Cardo, D.M., Cookson, S.T., Holmes, C.E.; Antunes, M.B., Melo Filho, D.A., Lyra, T.M., Barreto, V.S., Azevedo, S.M. & Jarvis, W.R. (1998). Liver failure and death after exposure to microcystins at a hemodialysis center in Brazil, *New England Journal of Medicine*, 338, pp. 873-878.

Knoll, M.F., Silva, A.C.F., Tedesco, S.B. & Canto-Dorow, T.S. (2006) Effects of *Pterocaulon polystachyum* DC. (Asteraceae) on onion (*Allium cepa*) root-tip cells. *Genet. Mol. Biol.*, 29, 3, pp. 539-542, *ISSN 1415-4757*

Laughinghouse IV, H.D. (2007). *Efeitos citotóxicos e genotóxicos de extratos aquosos de cepas de Microcystis aeruginosa (Chroococcales, Cyanobacteria)*, Thesis, Universidade de Santa Cruz do Sul – UNISC, Sant Cruz do Sul, RS.

Leme, D.M. & Marin-Morales, M.A. (2009). *Allium cepa* test in environmental monitoring: A review on its application, *Mutation Research*, 682, pp. 71-81, ISSN 0027-5107

Levan, A. (1938). The effect of colchicine on root mitoses in *Allium*, *Hereditas*, 24, pp. 471-486, ISSN 0018-0661.

Love, A. & Love, D (1975). *Plant chromosomes*. Lubrecht and Cramer Ltd, Monticello, NY.

Lubini, G., Fachinetto, J.M, Laughinghouse IV, H.D., Paranhos, J.T., Silva, A.C.F. & Tedesco, S.B. (2008). Extracts affecting mitotic division in root-tip meristematic cells, *Biologia*, 63, pp. 647-651.

Ma, T. H. (1999). The International Programme on plant bioassays and the report of the follow-up study after the hands-on workshop in China. *Mutation Research*, 426, pp. 103–106, ISSN 0027-5107

Ozmen, A. & Summer, S. (2004). Cytogenetic effects of kernel extracts from *Melia azedarach* L., *Caryologia*, 57, pp. 290–293.

Ragunathan I. & Panneerselvam, N. (2007). Antimutagenic potential of curcumin on chromosomal aberrations in *Allium cepa*, *Journal of Zhejiang University SCIENCE B*, 8, 7, pp. 470-475, ISSN 1673-1581

Rank, J. & Nielsen, M.H. (1994). Evaluation of the *Allium* anaphase-telophase test in relation to genotoxicity screening of industrial wastewater, *Mutation Research*, 312, 1, pp. 17-24, ISSN 0027-5107

Raven, H.P., Evert, R.E. & Eichorn, S.E. (1988). *Biologia Vegetal*. Guanabara Koogan, Rio de Janeiro

Rosa, S.G.T. & Ferreira, A.G. (2001). Germinação de sementes de plantas medicinais lenhosas, *Acta Botânica Brasílica*, 15, 2, pp. 147-54, ISSN 1677-941X

Rossato, L.V., Tedesco, S.B., Laughinghouse IV, H.D., Farias, J.G. & Nicoloso, F.T. (2010). Alterations in the mitotic index of *Allium cepa* induced by infusions of *Pluchea sagittalis* submitted to three different cultivation systems, *An. Acad. Bras. Ciênc.*, 82, 4, pp. 857-860, ISSN 0001-3765.

Schimidt, W., Willmitzer, H., Bornmann, K. & Pietsch, J. (2002). Production of drinking water from raw water containing cyanobacteria pilot plant studies for assessing the risk of microcystin break-through, *Environmental Toxicology*, 17, 4, pp. 375-385, ISSN 1522-7278

Sehgal, R., Roy, S. & Kumar, D.V.L. (2006). Evaluation of cytotoxic potential of latex of *Calotropis procera* and Podophyllotoxin in *Allum cepa* root model, *Biocell*, 30, 1, pp. 9-13, ISSN 1667-5746

Siddiqui, A.H., Tabrez, S. & Ahmad, M. (2011). Validation of plant based bioassays for the toxicity testing of Indian waters, *Environ Monit Assess*, 179, 1-4, pp. 241-253, ISSN 1573-2959.

Simões, C.M.O., Schenkel, E.P., Gossmann, G., Mello, J.C.P., Mentz, L.A. & Petrovick, P.R. (2001) *Farmacognosia: da planta ao medicamento*, (3 ed.), Ed. da UFSC, Florianópolis, SC.

Singh, R.J. (2002). *Plant Cytogenetics*. Boca Raton, CRC Press.

Silva, J. & Fonseca, M.B. (2003). Estudos Toxicológicos no Ambiente e na Saúde Humana, In: *Genética Toxicológica*, Silva, J., Erdtmann, B., Henriques, J. A. P. (Orgs.), pp. 69-84, Ed. Alcance, Porto Alegre.

Souza, L.F.B., Laughinghouse IV, H.D., Pastori, T., Tedesco, M., Kuhn, A.W., Canto-Dorow, T.S. & Tedesco,. S.B. (2010). Genotoxic potential of aqueous extracts of *Artemisia verlotorum* on the cell cycle of *Allium cepa*, *Int J Environ Stud*, 67, 6, pp. 871-877, ISSN 00207233

Vicentini, V.E.P., Camparoto, M.L., Teixeira, R.O. & Mantovani, M.S. (2001. *Averrhoa carambola* L., *Syzygium cumini* (L.) Skeels and *Cissus sicyoides* L.: medicinal herbal tea effects on vegetal and test systems. *Acta Scientiarum*, 23, 2, pp. 593-598, ISSN 1679-9283

Werner, V.R. & Laughinghouse IV, H.D. (2006) Cianobactérias: Qualidade da água e saúde pública. *Biologia atual*, 13, p. 2.

Teixeira, R.O., Camparoto, M.L., Mantovani, M.S. & Vicentini, V.E.P. (2003) Assesment of two medicinal plants *Psidium guajava* L. and *Achillea millefolium* L., in *in vitro* and *in vivo* assays, *Genetics and Molecular Biology*, 26, 4, pp. 551-555, ISSN 1415-4757

8

It's All in the Genes: How Genotype Can Impact Upon Response to Contaminant Exposure and the Implications for Biomonitoring in Aquatic Systems

Pann Pann Chung[1], Ross V. Hyne[2] and J. William O. Ballard[1]
[1]School of Biotechnology and Biomolecular Sciences, University of New South Wales
[2]Centre for Ecotoxicology
NSW Office of Environment and Heritage
Australia

1. Introduction

"Owing to this struggle for life, any variation, however slight and from whatever cause proceeding, if it be in any degree profitable to an individual of any species, in its infinitely complex relations to other organic beings and to external nature, will tend to the preservation of that individual, and will generally be inherited by its offspring."

– Charles Darwin, *On the origin of species by means of natural selection*

In 1859, Charles Darwin and Alfred Russell Wallace independently proposed the theory of evolution by natural selection. This theory propounded that the inter-generational success of all living organisms is restricted by their range of environmental tolerances and organismal requirements. In the current post-genomics era the impact of environmental stressors upon the *genetic structure* of populations and species continues to be a fascination from both theoretical and practical perspectives. From a theoretical perspective stressors can give deep insight into processes causing genetic subdivision. Practically, stressors can influence genetic diversity and this can provide us with information on the health of an ecosystem. While stressors in the environment can be both anthropogenic as well as natural in origin, in this review we focus on man-made stressors.

Stressors of anthropogenic origin can include physical modifications such as habitat alteration, biological agents such as introduced species, and chemical agents such as industrial pollutants. Of these, chemical pollution is arguably the stressor with the most immediate impact on the health of our biota. In the nearly three hundred years since the industrial revolution, environmental pollution has been on the increase, while the health of our environment has simultaneously suffered. Chemical pollution such as that from urban run-off, industrial discharge and pest control agents to name a few has been shown to have widespread detrimental and long-lasting effects on natural populations. Our aquatic environments are under particular threat as our rivers, estuaries, and oceans are often ecological sinks or dumping grounds for anthropogenic chemicals (e.g. Birch, 2000; Dunk et al., 2008; Jones, 2010).

Aquatic systems are the life-blood of many habitats, but are also subject to intense anthropogenic use for a variety of purposes including industry, transport as well as recreation. Thus, it is crucial to understand and monitor the health of aquatic organisms and aquatic ecosystems. For example, the British Petroleum (BP) *Deepwater Horizon* oil spill of April 2010 in the Gulf of Mexico saw approximately five million barrels-worth of crude oil lost into the ocean within the span of two months (Lubchenco et al., 2010), with an estimated 779 million litres spilt in total (Atlas & Hazan, 2011). Despite being the largest oil spill in United States history, several media reports suggested that based on observed wildlife mortalities the incident had only a moderate immediate environmental impact. It is likely, however, that immediate impact estimates represent less than 5% of the true biological effect (Williams et al., 2011). Incidences such as the BP oil spill highlight the immense gap in our knowledge about the extent and impact that anthropogenic stressors can have on our biota, especially in our aquatic environments.

The aim of this chapter is to discuss how genetic approaches linked with the tools of molecular biology can contribute to the fields of ecotoxicology and evolutionary toxicology. In the field of ecotoxicology, the importance of organismal genotype on the response of those organisms upon exposure to a contaminant has often been underexplored and its importance underestimated, particularly in the selection of suitable test organisms and sentinel species in biomonitoring programs. In this review, the term "biomonitoring" will be used as a collective representation of all terms relating to the detection and assessment of biological responses to environmental stressors. In the first section of this review we consider research on the effects of anthropogenic contaminants on the genetic structure of natural populations and then consider how the genotype of an organism can affect response to toxicant and contaminant exposure. In particular, we focus upon research that examines aquatic systems and aquatic sentinel species. We also explore the current genetic and molecular methodologies that are being utilised to assess the relationships between the underlying genetic structure of populations and the response to contaminant exposure. The focus of the second section of this chapter is traditional biomonitoring practices and the utilisation of sentinel species for biomonitoring. In this second section, we present a case study of ongoing research into the genetics and molecular biology of contaminant exposure response in an important local estuarine sentinel species in eastern Australia. We conclude that current molecular tools can assist in designing experimental and testing approaches to negotiate the influence of genotype in ecotoxicology.

2. Ecotoxicology and genetics

Ecotoxicology is the study of the effects of toxicants and contaminants, both natural and anthropogenic in origin, on the health of individuals and populations. Ecotoxicological studies involve the testing of model organisms with toxicants and contaminants of interest against reference populations or controls. Such studies are important for identifying substances that are potentially harmful to the health of both humans and other organisms, as well as determining the impact of toxicant and contaminant stressors upon natural ecosystems. The goal of this section is to highlight the need for objective methodologies to assess the impact of toxicant and contaminant exposure in biological systems.

It's All in the Genes: How Genotype Can Impact Upon Response to Contaminant Exposure and the Implications for
Biomonitoring in Aquatic Systems

133

In ecotoxicological studies, measurements of the effects of toxicants and contaminants on biological systems are typically estimated through organismal-level endpoints such as mortality, reproduction and growth. For example, reproduction and development of the ramshorn snail, *Planorbarius corneus*, is known to be impacted by exposure to sewage treatment effluent (Clarke et al., 2009). *P. corneus* is an egg-laying freshwater mollusc with a seasonal breeding pattern that is commonly found in European river systems. Snails collected from a clean waterway over two separate years were exposed to varying concentrations of effluent from sewage treatment facilities for 12-14 weeks in the laboratory and the reproductive output (measured as the number of eggs produced), egg quality (measured as egg mass), as well as gonad development analysed. Although exposure of *P. corneus* to sewage treatment effluent resulted in greater female fecundity with increased egg production among effluent-exposed animals, egg quality was found to be significantly decreased, suggesting a physiological trade-off between reproductive output and offspring viability (Clarke et al., 2009). Additionally, while the development of female gonadal tissue appeared unaffected by exposure to effluent, spermatogenesis in first generation offspring was found to be decreased; this is indicative of exposure to endocrine-disrupting chemicals common in many wastewater effluents (Clarke et al., 2009). While these observed differences in organismal endpoints are likely due to toxicant and contaminant exposure, results may also be confounded by seasonal environmental differences affecting the parental and grandparental generations in the sampling locality from which test organisms were derived. It has been demonstrated that individual fitness can also be affected by "carry-over effects" dating back to the parental and grandparental generations (Hercus & Hoffman, 2000), and this may have contributed to differences between the experimental groups in this study (Clarke et al., 2009).

A fundamental drawback of many ecotoxicological studies is that the results of toxicant and contaminant exposure analyses are indicative only of the health of the individuals being assessed rather than representing the response of the population or species as a whole (Bickham, 2011). In many studies large discrepancies are reported between individuals, test groups, treatments and replicates (e.g. Angus, 1983; Croteau & Luoma, 2009). This is particularly true for many behavioural traits such as exposure recovery or toxicant avoidance (Blomberg et al., 2003; Smith et al., 2010), making population-level and species-level inferences unreliable. For example, a test of phenol tolerance in natural populations of the mosquitofish *Gambusia affinis* identified differential tolerance between individuals (Angus, 1983). *G. affinis* were sampled from both industrially contaminated and clean rivers in the United States and exposed to a single concentration of phenol using a standard 48-hour toxicity assay, with the rate of mortality as the endpoint. The study identified differential tolerance to phenol among the test organisms, with some individuals resistant to phenol toxicity, some showing signs of resistance following acclimation in the presence of sublethal concentrations, while other individuals showed no resistance (Angus, 1983). This timeless study included wild-caught fish, which has unique experimental advantages and disadvantages. The primary advantage is that the results are directly relevant to the population tested. The disadvantage is that the observed differences in phenol sensitivity may be due to differences in the underlying genetics, age and health of the test organisms.

Life-history characteristics such as the developmental stage of the test organisms can also affect the observed response to toxicant exposure. This is particularly important if the overall health of an ecosystem is to be determined. For example, ecotoxicological studies frequently utilise organisms at specific life or developmental stages, such as embryos or juveniles, as they demonstrate greater sensitivity than mature organisms (Ansari et al., 2010; Hutchinson et al., 1998; Pollino et al., 2002). Thus, a life-history bias in environmental monitoring studies of sensitive taxa may indicate a contamination event. Mhadhbi et al. (2010) examined the effects of acute metal and polycyclic aromatic hydrocarbon (PAH) toxicity on the early life stages of a species of turbot fish, *Psetta maxima*. Fertilized eggs were exposed to a range of environmentally and biologically relevant concentrations of metals and PAHs over a six-day period, then exposure success as measured by embryo survival and developmental malformations in post-hatch fish larvae were compared. It was determined that embryos were more tolerant to metals and PAHs than hatched larvae, which showed pericardial oedema as well as skeletal abnormalities leading to deformed larvae (Mhadhbi et al., 2010).

In this section we have argued that ecotoxicological studies are important in identifying potentially harmful substances in the environment. Traditional ecotoxicological studies and biomonitoring programs typically make use of organismal-based endpoints such as percentage mortality or reproductive output to assess the impact of toxicant and contaminant exposure. However, these endpoints vary greatly between the populations assessed and this can result in data with poor reproducibility and large standard errors (Wirgin & Waldman, 2004). Moreover, toxicant and contaminant concentrations used in many standard toxicity studies are often necessarily high as to induce observable and measurable organismal-level effects. In the following section we show that differences in response to stressors of anthropogenic origin can be influenced by the underlying genetic makeup of the species, and this also needs to be controlled if ecotoxicological studies are to be robust and reproducible.

2.1 Genotype and differential tolerance to toxicant exposure

Differential genetic tolerance in response to stressors of anthropogenic origin has been demonstrated in a wide variety of organisms. This is not surprising. Most species and populations show heritable genetic variability that was arguably first noticed and discussed by Charles Darwin in 1859. What is perhaps more surprising is that the importance of Darwin's first two chapters has not been fully embraced in the broad cross-section of ecotoxicological studies. Perhaps this is because human mediated contamination arguably falls between Chapter 1 "Variation under domestication" and Chapter 2 "Variation under Nature". Low levels of contamination can change the gene frequencies in a population in the same way that sublethal concentrations of pesticides and insecticides cause resistance in insects (Garrett-Jones & Gramiccia, 1954; Livadas & Georgopoulos, 1953). In this section we demonstrate that the genetic makeup of populations and species influences response to stressors, and this needs to be controlled in ecotoxicological studies (Wirgin & Waldman, 2004).

In ecotoxicology, a common approach to examine the relationship between toxicants or stressors and organisms is to measure biologically significant endpoints such as post-exposure mortality, developmental processes, or reproductive output. Most often, the genetic structure of the organisms of interest is not examined. Yet this is likely to be important when documenting the environmental impact of a specific anthropogenic contamination event. Allelic frequencies within a population can, for example, be examined

It's All in the Genes: How Genotype Can Impact Upon Response to Contaminant Exposure and the Implications for
Biomonitoring in Aquatic Systems

135

in field studies and compared against control populations post hoc. For instance, the link between allozymatic genotype and differential survivorship has been demonstrated in the gammarid amphipod *Hyalella azteca* exposed to metals (Duan et al., 2001). One-month old laboratory cultured amphipods were exposed to acute concentrations of cadmium and zinc solutions in a 72-hour test. Mortality was measured as lack of response when the animal was probed; dead or moribund animals were periodically removed and preserved for subsequent allozymatic analysis, and remaining live animals were also sacrificed at the end of the test period. Exposure to both cadmium and zinc resulted in relatively high mortality; surviving cadmium-exposed *H. azteca* also displayed behavioural changes (hyperactivity) after 48 hours of exposure (Duan et al., 2001). Allozymatic analysis found that of the six genotypes identified, animals harbouring genotype AC had a significantly higher rate of survival than all other genotypes following cadmium exposure, whereas genotype CC displayed greater tolerance following zinc exposure (Duan et al., 2001). These findings also suggest differential modes of selection for different stressors, such that genotypes more resistant to a particular stressor may skew the genetic structure of the population in favour of those individuals harbouring the resistant allele.

A complementary experimental approach is to directly examine the relationship between specific genetic types and the response to toxicant and contaminant exposure. The knowledge that organisms harbouring different genotypes can demonstrate differential tolerance and susceptibility to toxicant and contaminant exposure has bearing upon the choice of test individuals for ecotoxicological studies. Studies have identified differential tolerance and susceptibility of specific genotypes to toxicant and contaminant exposure (Roelofs et al., 2009; Snyder & Hendricks, 1997). Differential survival across three different mitochondrial lineages has been demonstrated in the marine harpacticoid copepod *Microarthridion littorale* exposed to a mixture of pesticides (Schizas et al., 2001). Three naturally occurring mitochondrial lineages were previously identified and found to occur in approximately equal frequencies at a South Carolina reserve in the United States. *M. littorale* were sampled from this locality, and animals exposed to a pesticide mixture for a period of 24 hours. Survivorship was defined as those living individuals still displaying normal swimming ability. Following exposure, *M. littorale* were harvested and haplotyped to identify mitochondrial lineage. Of the three lineages identified and tested, it was found that lineage I displayed the highest rate of survival, significantly exceeding even that of the test controls, while lineage II showed greatest susceptibility to the pesticide mixture (Schizas et al., 2001). However, while the results of this study suggest a possible link between mitochondrial DNA and toxicant exposure response, they do not demonstrate direct causality; the role of nuclear DNA in toxicity response and the interaction between nuclear and mitochondrial genes still needs to be examined.

In this section, we have discussed how the genotype of an individual has significant bearing upon the potential response to toxicant and contaminant exposure. However, examining the impact of toxicant and contaminant exposure on organisms based on differences at single genetic loci can be considered the equivalent of measuring the water in the oceans one drop at a time. In the next section, we introduce the emerging field of *evolutionary toxicology* and review how this field of study is attempting to answer questions on how toxicants and contaminants are impacting the genetics of biological systems at the genomic and population levels.

2.2 Evolutionary toxicology

Here, we introduce the concept of evolutionary toxicology as an experimental strategy to investigate the influence of toxicant and contaminant exposure on natural populations. A central goal of evolutionary toxicology is to examine both the direct and indirect effects of toxicant and contaminant exposure on the genetic structure of natural populations (Bickham et al., 2000; Matson et al., 2006; Theodorakis et al., 2001). A chief advantage of evolutionary toxicology over traditional ecotoxicological studies is that the utilisation of molecular and population genetics methodologies means these studies are based on fundamental evolutionary concepts. As a consequence they are able to provide a more holistic representation of an entire population or species, rather than just specific individuals (Bickham, 2011).

Broadly speaking, direct effects (genotoxic effects) are those where exposure to the toxicant or contaminant cause direct damage to genetic material (e.g. insertion / deletion events, base changes, chromosomal rearrangements). Indirect (non-genotoxic) effects are non-mutagenic in nature but cause stress to exposed individuals and can result in downstream population-level effects such as altered reproductive success, survivorship or altered gene expression patterns (Bickham, 2011; Rose & Anderson, 2005).

The identification of a genetic basis for organismal and behavioural traits allows predictions to be made across the entire species and potentially between species. As an example, reburial behaviour is a normal stress response and hazard avoidance mechanism in many shellfish species, and is often indicative of organism health. An investigation into recovery behaviour following metal exposure in the New Zealand freshwater clam, *Sphaerium novaezelandiae* found that this behavioural response is genotype-dependent within this species (Phillips & Hickey, 2010).

Bickham (2011) divides the direct and indirect effects of toxicant and contaminant exposure into four key genetic outcomes which he refers to as the "four cornerstones of evolutionary toxicology". While these four cornerstones represent a useful basis for discussion they are by no means exhaustive of the causative effects of toxicant and contaminant exposure upon organisms. These four cornerstones as described by Bickham are:

1. contaminant-induced increases in mutation rates, resulting in changes in genetic frequencies in impacted populations (e.g. Matson et al., 2006; Theodorakis et al., 2001);
2. contaminant-induced selection at loci affecting survival success, resulting in changes in genetic frequencies in the population (e.g. Cohen, 2002; Theodorakis & Shugart, 1997);
3. alterations to patterns of genetic distribution and migration, leading to demographic shifts (e.g. Maes et al., 2005; Matson et al., 2006); and
4. genome-wide changes in genetic diversity (e.g. Armendariz et al., 2004; Connon et al., 2008; Poynton et al., 2008; for reviews on evolutionary toxicology, see: Bickham, 2011; Theodorakis & Wirgin, 2002; van Straalen & Timmermans, 2002).

In the following sections, we examine how these four cornerstones provide a practical guideline for the design of ecotoxicology and evolutionary toxicology studies, as well as review some of the molecular techniques currently being applied to research in each of these areas. A key advantage of molecular techniques in biomonitoring is that many of these techniques are more sensitive at detecting exposure effects, even at concentrations below the organismal-level effective concentrations (Poynton et al., 2008). The four genetic outcomes proposed are, however, by no means exhaustive, and novel techniques and new technologies will undoubtedly continue to further elucidate how toxicants and contaminants affect biological systems at the genetic level.

2.2.1 Contaminant-induced increases in mutation rates

The first cornerstone of evolutionary toxicology examines the effects of genotoxicant substances on the integrity of genetic material in biological systems (Bickham, 2011). Genotoxicants can increase mutation rates and directly interact with DNA, thereby causing heritable changes to the molecule. Hallmarks of exposure to genotoxicants can include increased base changes, insertion / deletion events, chromosomal rearrangements, or degradation of the DNA molecule (Rinner et al., 2011; Theodorakis et al., 1997, 2001). Patterns of mutation that typically arise from recent exposure events to genotoxicants are pockets of increased mutation rates restricted to individuals within an impacted area and which are not detected in unimpacted environments (Bickham, 2011). However, caution must be taken to ensure that the detected mutation events are the direct result of genotoxicant exposure rather than having arisen through spontaneous mutational events (Yauk et al., 2008).

Increased mutation rates have been correlated with regions of high industrial activity (e.g. Chung et al., 2008; Matson et al., 2005; Theodorakis et al., 1997). Rinner et al. (2011) identified increased rates of mutation associated with contaminant exposure in an invasive fish species in a highly contaminated region in Azerbaijan. The mosquitofish *Gambusia holbrooki* is an invasive fish species introduced to Azerbaijan in the 1930s to control mosquito densities and have since spread throughout the region. Given its relatively recent history in the region *G. holbrooki* populations are expected to be relatively homogeneous, with novel mutational events easily traceable (Rinner et al., 2011). In conjunction with the work by Matson et al. (2006), the genetic diversity at the mitochondrial control region was assessed in *G. holbrooki* sampled from the heavily contaminated Sumgayit region and surrounding reference sites. It was found that fish sampled from the contaminated region possessed four novel haplotypes and heteroplasmies, of which only one was found to occur among fish from reference populations. The other rare heteroplasmies identified were hypothesized to have been contaminant-induced mutational events as they were not found to occur outside of the contaminated region (Rinner et al., 2011).

Increased DNA damage (i.e. strand breakage) such as those induced by exposure to radioactive compounds can affect reproductive success and reduce reproductive fitness. Theodorakis et al. (1997) investigated the extent of DNA damage in the mosquitofish *G. affinis* exposed to radionuclides in the United States. Adult female fish were collected from liquid waste settling ponds as well as clean reference ponds, and liver tissue and blood samples analysed for DNA integrity using gel electrophoresis. In addition, the fecundity of sampled specimens was quantified in terms of embryo number and the occurrence of observable abnormalities. It was found that for both liver and blood samples, the occurrence of double-stranded DNA breakages was significantly higher in *G. affinis* collected from radionuclide-contaminated ponds than in fish sampled from reference ponds. The results were also reflected in both the percentages of abnormal broods and abnormal embryos counted in the sampled fish. Future studies interested in assaying cellular damage as a result of contaminant exposure may also quantify the concentrations of 8-oxodeoxyguanosine within a cell as it is a measure of oxidative stress (Nadja et al., 2001).

2.2.2 Contaminant-induced selection at loci affecting survival success

The second of the four cornerstones of evolutionary toxicology describes changes in genetic diversity or allelic frequencies within populations as a result of selection acting upon loci affecting the survivorship of organisms (Bickham, 2011). Exposure to toxicants or contaminants typically causes stress in organisms, which in turn can affect any number of downstream organismal endpoints including survivorship and reproductive success. In the words of Charles Darwin, "... *if variations useful to any organic being do occur, assuredly individuals thus characterised will have the best chance of being preserved in the struggle for life; and ... will tend to produce offspring similarly characterised*" (Darwin, 1859). Thus, individuals harbouring favourable genetic variations will likely be subject to positive selection, whereas those less tolerant are selected against, consequentially shifting the genetic structure of the population and / or species.

Where loci involved in the response to a particular stressor are known, it is possible to assess the effects of selection by quantifying allele frequencies or genetic diversity at those loci. Cohen (2002) examined the patterns of amino acid substitutions in the major histocompatibility complexes (*Mhc*) in the estuarine teleost fish *Fundulus heteroclitus* exposed to polychlorinated biphenyls (PCBs). *Mhc*s are a large class of proteins involved in vertebrate immune response by binding and presenting antigens to T-cells for further processing. High genetic variability in *Mhc* genes is maintained by pathogen and antigen-driven balancing selection (Robinson et al., 2003). In this study, direct sequencing of the *Mhc* class II B genes was performed and referenced against the mitochondrial hypervariable control region. It was found that population-specific amino acid replacements were correlated with contaminant exposure (Cohen, 2002). Additionally, *F. heteroclitus* sampled from PCB contaminated sites and clean reference localities displayed habitat-specific *Mhc* patterns that were not reflected in the sequences of the mitochondrial reference region (Cohen, 2002). *F. heteroclitus* sampled from the contaminated sites also showed significantly elevated rates of loci-specific non-synonymous amino acid substitutions that were not detected in fish from control sites. Together, these differences in substitution patterns suggest a selective stress response to PCB exposure in *Mhc* proteins (Cohen, 2002).

A major hurdle to detecting selection occurs when the stressor is unknown, or when the specific stress response loci or pathways have not been identified. One approach to overcome this "needle in a haystack" problem is to assay the cellular bioenergetics of impacted organisms. Stress response is often an energetically costly process, whereby cells must either bear toxicant accumulation load costs or induce cellular excretory mechanisms to clear the presence of the stressor. Mitochondria are responsible for the majority of ATP production in the cell, and thus measuring mitochondrial bioenergetics can be indicative of exposure to toxicant or contaminant stress. Cells undergoing stress response may either up-regulate ATP production to cope with the presence of the stressor, or toxicant exposure may inhibit mitochondrial function. Mitochondrial bioenergetics can be measured in a number of ways, including oxygen consumption and transmembrane potential (da Silva et al., 1998; Toro et al., 2003; Vijayavel et al., 2007). For example, Toro et al. (2003) assessed the effects of exposure to organic pollutants on bioenergetic responses in the giant mussels, *Choromytilus chorus* sampled from off the coast of Chile. *C. chorus* were assayed for contaminant clearance rate, oxygen consumption, and scope for growth, which is indicative of the energy budget dedicated to growth and reproduction. The study found that animals sampled from the heavily polluted locality demonstrated both lower rates of contaminant clearance from tissues, higher rates of oxygen consumption, as well as a negative energy budget for growth and reproduction (Toro et al., 2003). On the other hand, *C. chorus* from the unimpacted site showed higher clearance rates and lower oxygen requirement, with a strongly positive scope for growth, thus indicating high energy expenditure associated with toxicant exposure response (Toro et al., 2003). An alternative, or perhaps complementary, approach would be the measurement of metabolic rates in specific tissues such as permeabilised muscle fibres using a respirometer such as the OXYGRAPH-2K (Oroboros Instruments, Innsbruck, Austria) (Pichaud et al., 2011). Contaminants and toxicants can have tissue-specific effects; thus the measurement of metabolic efficiency in mitochondria-rich tissues such as muscle fibres allows for targeted assessment of the effect of environmental contaminant exposure (Pichaud et al., 2011).

2.2.3 Alterations to patterns of genetic distribution and migration

The third cornerstone of evolutionary toxicology examines the effect of stressors on gene flow and migration that lead to changes to inter-population genetic structure (Bickham, 2011). Migration and gene flow between populations are fundamental to the demographic and evolutionary stability of populations, particularly in cases where population sizes are small. Stressors such as anthropogenic toxicants or contaminants can affect the direction of movement between populations and hence the distribution of genotypes and allele frequencies (Maes et al., 2005; Matson et al., 2006).

Assessment of genetic diversity through functional enzymatic assays can be indicative of the fitness costs incurred by toxicant and contaminant stressors. Maes et al. (2005) assayed allozymatic and microsatellite genetic diversity in the catadromous yellow eel *Anguilla anguilla* from metal contaminated rivers in Belgium, where populations have been on the decline over the past two decades. Adult *A. anguilla* were sampled from three interconnected contaminated river basins, and muscle and liver tissue assayed for metal accumulation and genetic diversity. *A. anguilla* collected from the most polluted river basin showed the highest levels of metal accumulation and showed the least heterozygosity at all allozymatic loci investigated, which was not reflected in the microsatellite analysis. This suggests a genotypic shift toward specific homozygote classes and / or a greater ability to clear metal accumulation by homozygotes in the most contaminated system (Maes et al., 2005). Future studies may also consider how mitochondrial gene expression changes with toxicant exposure as upregulation of genes encoding mitochondrial subunits has been shown to occur with mild mitochondrial dysfunction (Ballard et al., 2010).

Population genetics methodologies have also been applied to assess genetic diversity and patterns of gene flow between populations. Matson et al. (2006) examined the effects of chronic chemical contaminant exposure on population genetic structure and genetic diversity in marsh frogs from Azerbaijan. *Rana ridibunda* were collected from several localities within a region of known contamination as well as from two clean reference sites surrounding this region. Genetic diversity at the mitochondrial control region was measured and the genetic distances between populations calculated to determine migration patterns between populations. This study identified a significant difference in the level of genetic diversity between the contaminated and reference regions as indicated by both haplotypic and nucleotide diversity, where overall mitochondrial diversity among *R. ridibunda* from the contaminated region was significantly lower (Matson et al., 2006). Analysis of genetic differentiation and gene flow between populations indicated that the rate of migration into the contaminated region far exceeded that of outward movement, suggesting *R. ridibunda* in the contaminated region have both decreased fitness and reproductive success, and that inward migration provided a compensatory effect for decreased reproductive success of the impacted population.

2.2.4 Genome-wide changes in genetic diversity

In recent years, genome-wide "omic" approaches have become increasingly commonplace in evolutionary toxicology (Snape et al., 2004; Van Aggelen et al., 2010). Omic approaches encompass whole genome sequencing studies, transcriptomic studies, as well as proteomic expression studies, with the latter two being most commonly applied in ecotoxicology and evolutionary toxicology (e.g. Iguchi et al., 2007; Poynton et al., 2008; Zhou et al., 2010). Although some toxicants may impact exclusively upon specific loci or cellular machinery, organismal response and recovery to toxicant exposure is likely to be a genome-wide process; hence, the ability to encompass and measure total cellular response to toxicant exposure is crucial in evolutionary toxicology (Bickham, 2011). An additional characteristic of both transcriptomic and proteomic profiling worth considering is that results are not only reflective of the underlying genetic structure of the organism, but are also indicative of the functionality of differences in the genetic structure, such that non-functional genetic changes (e.g. silent mutations in non-coding regions) are generally not detected. Thus, it can be considered that transcriptomic and proteomic analyses may be able to better identify biologically significant genetic changes induced by toxicant exposure.

Transcriptomic analyses examine genome-wide gene expression and are commonly used to compare differences in transcript profiles between organisms exposed and unexposed to toxicants and contaminants. A further application of transcriptomics in biomonitoring is the potential to identify novel informative genetic markers of exposure by pinpointing genes and loci that are differentially expressed. The use of transcriptomics to create toxicant exposure expression profiles and for biomarker discovery has been well documented in the widely used model organism *Daphnia magna* (e.g. Vandegehuchte et al., 2010; Watanabe et al., 2008). *D. magna* has demonstrated high sensitivity to a wide range of toxicants and contaminants, and distinct transcriptomic profiles have been generated for a number of metals and organic toxicants such as copper, zinc, lead and PAHs (Poynton et al., 2008). A major advantage of transcriptomics lies in the high detection sensitivity of these assays. Unlike standard toxicology tests that often require high doses of toxicants to induce a measurable effect in the test organism, gene expression patterns can differ significantly even with exposures to very low toxicant concentrations. In fact, distinct expression differences have been detected at concentrations as low as one-twentieth of the 50% effective concentration (EC50) value (Poynton et al., 2008). Moreover, it was found that higher exposure concentrations often resulted in the loss of toxicant-specific expression patterns in favour of a generalised stress responses (Poynton et al., 2008), making transcriptomic profiling a powerful and sensitive tool for biomonitoring. However, in addition to the high cost a disadvantage of transcriptomics is the presence of transcriptional "noise" as a result of stochastic gene expression, and these random differences in transcript levels between samples and even individual cells must necessarily be accounted for when conducting transcriptomic analyses, particularly in the case of low abundance transcripts (Raj & van Oudenaarden, 2008).

Similarly, proteomic studies have been utilised to generate genome-wide protein expression profiles. Zhou et al. (2010) conducted protein expression assays on farm-reared abalone *Haliotis diversicolor supertexta* exposed to select endocrine disrupting compounds. *H. diversicolor supertexta* were exposed to low concentrations of two plasticiser compounds (1% of the previously established 50% lethal concentration (LC50) value) and hepatopancreatic tissue assayed for protein expression differences. The study identified 27-35 differentially expressed proteins in *H. diversicolor supertexta* exposed to the plasticisers when compared against the untreated control; nearly 20 of the identified proteins were up-regulated proteins, and 9-16 proteins were down-regulated (Zhou et al., 2010). Mass spectrometry identified many of the up-regulated proteins to be associated with general stress response, while several of the down-regulated proteins were involved in metabolism. In addition, the study also identified a number of compound-specific differentially expressed proteins.

3. Environmental monitoring

In this section we discuss the advantages of environmental monitoring to determine the health of natural systems and then consider the use of an amphipod crustacean to monitor the health of estuarine sediments in eastern Australian waterways. Environmental monitoring or detecting the presence of toxicants and contaminants falls under the two broad categories of *bioindication* and *biomonitoring*. In this section we focus on the latter.

Bioindication is a qualitative assessment of the impact of environmental stressors (Holt & Miller, 2011). An example is the destabilisation of lysosomes in the oyster, *Crassostrea virginica* exposed to environmental stressors (Ringwood et al., 1998). Lysosomes are considered to be sensitive organelles to contaminant-induced stress as they are involved in a host of cellular repair and defence functions. In this study, *C. virginica* exposed to high salinities as well as copper solutions displayed higher levels of lysosomal destabilisation than control organisms. These results were also congruent with field studies of *C. virginica* deployed in contaminated estuarine sediments (Ringwood et al., 1998).

Biomonitoring is the process of quantitatively determining the impact of environmental stressors on biological systems (Holt & Miller, 2011). Kreitsberg et al. (2010) measured the effects of PAH exposure in flounder following an oil spill off the coast of Estonia. In this study, *Platichthys flesus trachurus* were sampled at two-month intervals beginning five months post-spill and liver samples taken to examine PAH accumulation. Samples were analysed for PAH concentrations by high-performance liquid chromatography. Biologically high concentrations of PAHs were detected in the liver tissue of *P. flesus trachurus* caught five months after the spill, which was found to decrease over time (Kreitsberg et al., 2010).

The chief purpose of biomonitoring is the measurement and tracking of chemical substances in biological systems with the purpose of monitoring and assessing exposure to stressors, particularly those of anthropogenic origin. The assessment of toxicants and contaminants in biological systems through biomonitoring provides an environmentally relevant indication of the impact of environmental stressors within a given habitat or ecosystem. Direct quantification of contaminants such as sediment and water using chemistry-based ecotoxicological methodologies may not give any indication of biological or ecological relevance. Furthermore, chemical and physical analyses are also unable to assess ecological processes such as population shifts, nor assess levels of biodiversity within an ecosystem.

Biomonitoring can be conducted on both whole organisms as well as biological materials such as specific tissues, bodily fluids or cells. In biomonitoring, assessing the effects of toxicants and contaminants on whole organisms is most common; in this review we focus on environmental biomonitoring. Traditionally, two broad organismal measures of sensitivity or tolerance of biota toward toxicants and contaminants are commonly applied in biomonitoring (Mandaville, 2002). One is a measure of organism sensitivity, and the second is a snapshot assessment of the quality of a particular environment. A major shortcoming of both traditional biomonitoring measures is the underlying assumption that related organisms will demonstrate similar responses toward toxicant exposure. This is not always true. It has been documented that sister species can show differential susceptibility to toxicants and contaminants (e.g. King et al., 2004, 2006a).

In the following section, we examine the use of sentinel species in biomonitoring. We then consider the use of an amphipod crustacean in biomonitoring in eastern Australia.

3.1 Sentinel species in biomonitoring

As with ecotoxicological studies, biomonitoring studies and programs typically involve assessing the response of model organisms or *sentinel species* (also known as bioindicators or biomonitors) to toxicant and contaminant exposure. Perhaps the earliest and best known example of a sentinel species is the canary in the coal mine. Even as recently as the late 20th century, it was customary for coal miners to take canaries deep into the mine shafts with them as an early warning system against the build-up of toxic gases. The birds, being more sensitive to the presence of methane and carbon monoxide gas than humans, would fall ill and perish long before gas concentrations became dangerous for the miners, allowing workers to escape the tunnels safely. Environmental sentinel species serve the same function as the canary in the coal mine, providing an early detection system to the presence and effects of toxicants and contaminants in the environment.

A sentinel species can be any ecologically relevant organism that is well characterised, locally abundant and simple to survey and culture. Basic criteria of a good sentinel species are that it should be sensitive to the toxicant or contaminant of interest at both biologically and ecologically relevant concentrations, and do so in a reproducible manner that is reflective of the whole population, species and / or ecosystem. A good sentinel species should also possess readily measurable responses that may include reproductive output, developmental check-points or measures of growth (Holt & Miller, 2011). A limitation of many sentinel species is that the intraspecific variation in the response to a contaminant is not known because the underlying genetic variation in that species has not been quantified. This is of particular concern when a sentinel species is exposed to a range of contaminants such that distinct genetic pathways are differentially affected. In the extreme case, a specific species may be a highly sensitive sentinel with regard to one contaminant but a less sensitive sentinel with regard to a second contaminant.

The use of sentinel species to assess the impact of environmental stressors can give biological insight. Sentinel species essentially serve three major functions:

1. to assist in monitoring changes in the physical and / or chemical environment that can be of either natural (e.g. temperature, precipitation, salinity) or anthropogenic origin (e.g. use of chemical agents, habitat restructuring, industrial contamination);
2. to assist in tracking ecological processes (e.g. changes in population densities); and
3. monitoring the biodiversity of a given habitat or ecosystem (Holt & Miller, 2011).

In the following section, we present a case study of ongoing research applying some of the above methodologies to a sentinel species employed in eastern Australia to assess the health of estuarine sediments and ecosystems.

3.2 Case study: *Melita plumulosa*, the environmental canary of Eastern Australian waterways

In many aquatic systems, benthic organisms such as crustaceans serve as the "environmental canaries" of aquatic environments. These invertebrate species possess many useful qualities that make them ideal sentinel species to monitor the health of waterways. Among the macroscopic bottom-level feeders in a food web, benthic organisms are some of the most abundant members of an ecosystem and facilitate the transfer of energy and nutrients between microbial communities and the higher members of the food web. Most species of aquatic invertebrates feed by filtering or ingesting nutrients from the water or sediment, and thus can accumulate high concentrations of toxicants or contaminants that may be present within a habitat. It has been observed that population disruption or loss of bottom-level organisms and benthic invertebrate species can result in the disruption and even collapse of entire ecosystems (Owens & Dittman, 2003; Pillay et al., 2010; Wallace & Webster, 1996).

Among aquatic invertebrate species, crustaceans of the order Amphipoda are some of the most well characterised and commonly utilised organisms in biomonitoring studies. Amphipods are a highly diverse order of crustaceans found to inhabit nearly all marine and freshwater habitats worldwide and are a major benthic component in terms of both biomass and species diversity. To date, nearly one hundred amphipod families have been described (Lowry et al., 2000). Because of their ecological importance, numerical abundance as well as sensitivity to a variety of anthropogenic toxicants and contaminants, amphipods have long been utilised as sensitive environmental sentinels (e.g. Castro et al., 2006; Roach et al., 2001; Wu & Or, 2005). In particular, amphipods are often utilised to monitor the health of aquatic sediments, which can serve as an ecological sink for many anthropogenic toxicants and contaminants (Chapman & Wang, 2001; De Lange et al., 2006; Simpson et al., 2005). Other aquatic sentinel species include fish, frogs, molluscs, copepods and water fleas (e.g. Matson et al., 2006; Poynton et al., 2008; Rinner et al., 2011; Ringwood et al., 1998; Schizas et al., 2001).

3.2.1 Biology of *Melita plumulosa*

In eastern Australia, the local amphipod species, *Melita plumulosa* (Zeidler's Melita; family: Melitidae) (Fig. 1) is a commonly utilised sentinel species to assess the health of estuarine sediments (Ecotox Services Australasia, 2009). Originally isolated and described by Zeidler from specimens found in a coastal pool 100 m from the sea in northern New South Wales (Zeidler, 1989), M. plumulosa is an epibenthic amphipod endemic to eastern Australian waterways that can be found in the intertidal zone beneath rocks or shell-grit (King et al., 2006b; Lowry et al., 2000). Although the full geographic range of this amphipod has not been definitively established, historically it has been found in many waterways and estuaries between Brisbane and Melbourne along the east coast of Australia (Lowry et al., 2000); recently this amphipod has been sampled from the Pine River in Queensland down to the Yarra River in Victoria (Fig. 2).

In many river systems, M. plumulosa are found to co-localise with a sister species, Melita matilda (Hyne et al., 2005). In the field, M. plumulosa and M. matilda appear to cohabit the same environmental niches and share similar morphologies. Under the microscope, these two species can be definitively distinguished by an additional posterodorsal spine on urosomite 1 that is unique to M. plumulosa (Fig. 1) (Lowry et al., 2000). Although these two organisms are often found to co-localise and commonly sampled together, a study testing the sensitivities of eight local amphipod species in Australia and New Zealand to anthropogenic toxicants and contaminants demonstrated significantly different sensitivities between these two sister species (King et al., 2006a). Therefore, M. plumulosa has been designated as the sentinel species for monitoring sediment contamination in eastern Australia.

M. plumulosa and can tolerate a wide range of environmental conditions, including sediment particle sizes ranging from silt to gravel, salinities ranging from freshwater to seawater, as well as a wide range of temperatures (Hyne et al., 2005; King et al., 2006b). Under laboratory conditions, the average lifespan of M. plumulosa is 8-11 months; gravid females release live young that become sexually mature in four to six weeks (Hyne et al., 2005). Males are readily distinguishable from females by their larger gnathopod 2 as well as having more bristles along the antennae (Hyne et al., 2005). Optimal culture conditions for this amphipod as established by Hyne et al. (2005) are seawater of 25‰ salinity at 25°C ambient temperature on sediment composed of > 96% silt.

The advantages of utilising M. plumulosa as a sentinel species are manifold. The key advantage to employing this species for biomonitoring is the sensitivity of this amphipod to a range of sediment-bound contaminants (King et al., 2006a). Many contaminants often bind to and accumulate in aquatic sediments (Birch, 2000), the concentrations of which may not be reflected in the overlying pore water and water column. As this amphipod displays epibenthic behaviour and feeds on detritus by ingesting the sediment (Hyne et al., 2005), it has been demonstrated that M. plumulosa is particularly sensitive to sediment-bound contaminants (King et al., 2005, 2006b). Additionally, laboratory cultures of M. plumulosa are relatively simple to establish and maintain, as this amphipod has demonstrated robustness to a wide range of environmental conditions, and the optimal culture conditions have been well established (Hyne et al., 2005).

A disadvantage of this organism is the undetermined distribution of this amphipod. Although M. plumulosa has been found in many estuaries along the eastern Australian coast (Fig. 2), its established geographic distribution is not global across Australia and is predominantly limited to brackish sediments along the coastal fringe (Lowry et al., 2000). Furthermore, where M. plumulosa and M. matilda co-localise and in mixed-species laboratory cultures, M. matilda often outnumber and outcompete M. plumulosa (Hyne et al., 2005), making sampling time-consuming and correct identification crucial. Therefore, the applicability of this organism as a global sentinel species across Australian waterways is limited. A second disadvantage of *M. plumulosa* is the variability in female fecundity (Gale et al., 2006; Mann & Hyne, 2008). Female fecundity is a commonly assessed organismal endpoint of toxicant exposure in *M. plumulosa*, and is closely linked with dietary fatty acid composition which is crucial to ovary maturation and embryo development in amphipods and other crustacea (Clarke et al., 1985; Hyne et al., 2009; Middleditch et al., 1980). Recently, fine-milled silica has been shown to be a feasible alternative standardised substrate for short term toxicity tests; however, field collected sediments are still necessary for both long term tests and laboratory culture maintenance (Mann et al., 2011).

Fig. 1. The estuarine sentinel species of eastern Australia, the amphipod *Melita plumulosa*. The distinguishing morphological feature of this species is the additional posterodorsal spine on urosomite 1 (circled in red).

3.2.2 Ecotoxicology of *Melita plumulosa*

A key characteristic that makes the amphipod *M. plumulosa* an ideal sentinel species for monitoring and assessing sediment health is that it feeds on detritus by ingesting sediment, and is therefore directly exposed to sediment-bound toxicants and contaminants (Hyne et al., 2005; King et al., 2005; Spadaro et al., 2008). Studies have demonstrated this amphipod to be sensitive to both aqueous and sediment-bound metals and PAHs in both acute and chronic exposures, with juvenile animals displaying greater sensitivity than adult *M. plumulosa* (Gale et al., 2006; Hyne et al., 2005; King et al., 2005, 2006; Spadaro et al., 2008).

A range of toxicology tests have been established using *M. plumulosa*. These include a 42-day (full life-cycle) chronic whole-sediment test using metal-spiked sediments (Gale et al., 2006), a 10-day acute whole-sediment test using both adult and juvenile animals and metal-spiked sediments (King et al., 2006b), and more recently a 13-day reproduction test using both laboratory cultured amphipods as well as *in situ* testing (Mann et al., 2009, 2010, 2011). In all cases genetically heterogeneous amphipods were employed, with large numbers of individuals needed to offset the potential effects of inter-individual variation in toxicant

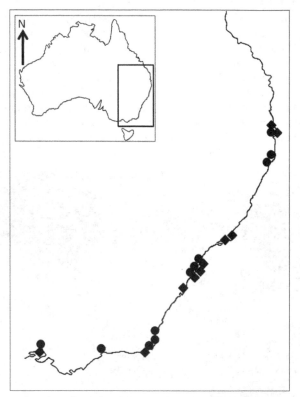

Fig. 2. Known geographic distribution of *Melita plumulosa* along the eastern coast of Australia. Historic sampling sites (1989 – 2000) are represented by circles, and current sampling sites (2000 – present) are represented by diamonds (Historic data courtesy J.K. Lowry and current data courtesy R.M. Mann).

response. Even with the vast numbers of individuals utilised in each of these tests (in excess of 200 and up to 1 000 individuals for the reproduction test (Mann et al., 2009)), a large degree of variance can still be observed both within and between replicates and treatments. Previously, we demonstrated that different genotypes can give rise to differential stress response; therefore, one possible explanation for the large discrepancies observed may be attributed to the use of genetically heterogeneous test populations. An alternate explanation for these results is the differences in age of test animals and hence reproductive potential of the individuals used.

It has also been demonstrated that the results of laboratory tests on *M. plumulosa* are not always replicable in *in situ* bioassays (Mann et al., 2010; Wilkie et al., 2010). A study by Mann et al. (2010) showed that laboratory tests demonstrated the contaminated field sediment to have high reproductive toxicity and confirmed previously established results; however, amphipods used in *in situ* bioassays displayed no evidence of reduced reproductive success. It was hypothesised that these discrepancies between the laboratory test results and the *in situ* bioassays are likely due to tidal water exchanges that can affect the bioavailability of contaminants in the overlying water as compared to the static laboratory tests (Mann et al., 2010). Additionally, other environmental fluctuations such as precipitation and the presence of other biota not present in the laboratory tests may also contribute to the bioavailability of contaminants within each test system.

The sensitivity of this amphipod to a range of toxicants and anthropogenic contaminants is not disputed. However, ecotoxicological studies such as these highlight the clear need for methodologies and testing protocols that include an assessment of how genetic variation within a species influences contaminant response. In the following section we briefly discuss what is currently known about the genetics of *M. plumulosa*.

3.2.3 Genetics of *Melita plumulosa*

Over the past decade, the biology and ecology of the amphipod *M. plumulosa* has been well characterised, and much has been established regarding its organismal-level response to toxicant and contaminant exposure. However, there remains a large gap in the knowledge of the genetics of this important sentinel species.

Currently, ongoing research into the population genetics as well as the genetics of the toxicological response in *M. plumulosa* is being conducted. For example, a study of the effects of chronic contaminant exposure on natural populations of this amphipod in eastern Australia identified genetic subdivision between impacted and reference populations, as well as significant life-history trait differences between the two populations (Chung et al., 2008). In this study, amphipods were sampled from one industrially contaminated waterway and one clean reference waterway, and the genetic diversity at one mitochondrial and one nuclear locus assessed. In addition, amphipod size and female fecundity as measured by the number of embryos per gravid female were also determined to assess whether the genetic and organismal data were correlative. The results of this study identified discrepancies in the level of genetic diversity between the two genetic loci examined, with mitochondrial genetic diversity notably higher among amphipods sampled from the contaminated river system than those from the reference waterway; this pattern of genetic diversity was not reflected at the nuclear locus (Chung et al., 2008). Furthermore, life-history trait variations also identified amphipods sampled from the contaminated river as being significantly smaller in size and less fecund than those sampled from the reference waterway, suggesting that chronic exposure to industrial contaminants has impacted both the health and genetic structure of the local amphipod population (Chung et al., 2008). A major limitation of this study is that amphipods from only two waterways were examined; analysis of additional populations and waterways is necessary to determine if the observed trends are indeed indicative of toxicant exposure, or if these data represent an isolated phenomenon.

Further research into the underlying genetic structure of *M. plumulosa* is currently underway, including an assessment of responses to other industrial contaminants prevalent in contaminated waterways in eastern Australia. To address the genetic concerns raised in this chapter, current research on this sentinel species involves the development and use of animals of standardised genetic lineage for toxicity assays. Furthermore, whole-genome assays as well as bioenergetic analyses have also been proposed.

4. Conclusion

The impact of anthropogenic toxicants and contaminants on the health of natural populations is an evolutionarily recent phenomenon given the brief industrial history of the human race. However, even though the footprints of anthropogenic toxicants and contaminants are relatively fresh, the effects upon biota have been demonstrated to be persistent and severe. Perhaps one saving grace of mankind is our ability and our desire to investigate and understand our impact upon the natural world.

In this chapter we reviewed past and current research on the effects of toxicants and contaminants on the genetic structure of natural populations, with a particular focus on aquatic, estuarine and marine systems. Genetics underpins all biological processes and it is of crucial importance that we understand the impacts that anthropogenic stressors have upon organisms at the molecular level. While it is clear that in this post-genomics era many powerful and informative tools are available to assess the impact of anthropogenic stressors on natural systems, further research must still be undertaken to complete our understanding of the consequences of toxicant and contaminant exposure on natural populations and species.

The four cornerstones of evolutionary toxicology proposed by Bickham (2011) provide a descriptive and practical framework by which the genetic-level effects of toxicant and contaminant exposure can be classified. Furthermore, this framework provides ecotoxicologists and evolutionary toxicologists with a practical and generalised guideline by which the problem of understanding the mechanisms of how anthropogenic stressors impact upon biological systems can be approached.

While we continue to expand our understanding of the mechanisms and impacts anthropogenic stressors have upon biota, biomonitoring is also crucial to keep track of the damage to natural populations occurring in our present. Given our current understanding of genetics and molecular processes, a key consideration is the effect of genotype on an organism's response to stressors. In the past, biomonitoring techniques have typically taken a broad-brush approach because there was an assumption that the mean response of an organism was more or less similar between individuals, as well as being sufficiently representative of that of the total population or species. However, it is now well understood and accepted that there is a significant influence of genotype on individual stress response variations, and this fact cannot be ignored.

Current molecular tools can assist in designing experimental and testing approaches to negotiate the influence of genotype. These may include conducting toxicology tests according to genotype and directly assessing the genetic impact of toxicants and contaminants. Additionally, molecular tools can help us identify the most appropriate sentinel species and individuals to assess the impact of toxicant and contaminant exposure. Nevertheless, it must be remembered that the genome of any organism is a large and often poorly understood minefield with complex molecular interactions. As a consequence it is unlikely that any one methodology will be able to fully account for all the minute levels of variation and interactions within biological systems.

5. Acknowledgments

The authors wish to thank J.K. Lowry for the historic data on *M. plumulosa* sampling localities. R.M. Mann is thanked for the current data on *M. plumulosa* sampling localities and his support with current work. We also wish to thank S.E. Hook and L. Tremblay for reviewing this chapter, and K.M. Cairns for commenting on early drafts.

6. References

Angus, R.A. (1983). Phenol tolerance in populations of mosquitofish from polluted and nonpolluted waters. *Transactions of the American Fisheries Society*, Vol. 112, No. 6, pp. 794-799, ISSN 0002-8487

Ansari, Z.A., Farshchi, P. & Faniband, M. (2010). Naphthalene induced activities on growth, respiratory metabolism and biochemical composition in juveniles of *Metapenaeus affinis* (H.Milne Edward, 1837). *Indian Journal of Marine Sciences*, Vol. 39, No. 2, (June 2010), pp. 285-289, ISSN 0379-5136

Armendariz, A.D., Gonzalez, M., Loguinov, A.V. & Vulpe, C.D. (2004). Gene expression profiling in chronic copper overload reveals upregulation of *Prnp* and *App*. *Physiological Genomics*, Vol. 20, No. 1, (December 2004), pp. 45-54, ISSN 1094-8341

Atlas, R.M. & Hazen, T.C. (2011). Oil biodegradation and bioremediation: A tale of the two worst spills in U.S. history. *Environmental Science and Technology*, Vol. 45, No. 16, (August 2011), pp. 6709-6715, ISSN 0013-936X

Ballard, J.W.O, Melvin, R.G., Lazarou, M., Clissold, F.J. & Simpson, S.J. (2010). Cost of a naturally occurring two-amino acid deletion in cytochrome *c* oxidase subunit 7A in *Drosophila simulans*. *The American Naturalist*, Vol. 176, No. 4 (October 2010), pp. E98-E108, ISSN 0003-0147.

Bickham, J.W., Sandhu, S.S., Hebert, P.D.N., Chikhi, L. & Anthwal, R. (2000). Effects of chemical contaminants on genetic diversity in natural populations: Implications for biomonitoring and ecotoxicology. *Mutations Research*, Vol. 463, No. 1, (July 2000), pp. 33-51, ISSN 1383-5742

Bickham, J.W. (2011). The four cornerstones of Evolutionary Toxicology. *Ecotoxicology*, Vol. 20, No. 3, (May 2011), pp. 497-502, ISSN 0963-9292

Birch, G.F. (2000). Marine pollution in Australia, with special emphasis on central New South Wales estuaries and adjacent continental margin. *International Journal of Environmental Pollution*, Vol. 13, No. 1-6, (2000), pp. 573-607, ISSN 0957-4352

Blomberg, S.P., Garland, T. & Ives, A.R. (2003). Testing for phlyogenetic signal in comparative data: Behavioral traits are more labile. *Evolution*, Vol. 57, No.4, (April 2003), pp. 717-745, ISSN 1558-5646

Castro, H., Ramalheira, F., Quintino, V. & Rodrigues, A.M. (2006). Amphipod acute and chronic sediment toxicity assessment in estuarine environmental monitoring: an example from Ria de Aveiro, NW Portugal. *Marine Pollution Bulletin*, Vol. 53, No. 1-4, (April 2006), pp. 91-99, ISSN 0025-326X

Clarke, A., Skadsheim, A. & Holmes, L.J. (1985). Lipid biochemistry and reproductive biology in two species of Gammaridae (Crustacea: Amphipoda). *Marine Biology*, Vol. 88, No. 3 (September 1985), pp. 247-263, ISSN 0025-3162.

Clarke, N., Routledge, E.J., Garner, A., Casey, D., Benstead, R., Walker, D., Watermann, B., Gnass, K., Thomsen, A. & Jobling, S. (2009). Exposure to treated sewage effluent disrupts reproduction and development in the seasonally breeding ramshorn snail (subclass: Pulmonata, *Planobarius corneus*). *Environmental Science and Technology*, Vol. 43, No. 6, (March 2009), pp. 2092-2098, ISSN 0013-936X

Chapman, P.M. & Wang, F. (2001). Assessing sediment contamination in estuaries. *Environmental Toxicology and Chemistry*, Vol. 20, No. 1 (January 2001), pp. 3-22, ISSN 1552-8618

Chung, P.P., Hyne, R.V., Mann, R.M. & Ballard, J.W.O. (2008). Genetic and life-history trait variation of the amphipod *Melita plumulosa* from polluted and unpolluted waterways in eastern Australia. *Science of the Total Environment*, Vol. 403, No. 1-3, (September 2008), pp. 222-229, ISSN 0048-9697

Cohen, S. (2002). Strong positive selection and habitat-specific amino acid substitution patterns in MHC from an estuarine fish under intense pollution stress. *Molecular Biology and Evolution*, Vol. 19, No. 11, (November 2002), pp. 1870-1880, ISSN 0737-4038

Connon, R., Hooper, H.L., Sibly, R.M., Lim, F.-L., Heckmann,L.-H., Moore, D.J., Watanabe, H., Soetaert, A., Cook, K., Maund, S.J., Hutchinson, T.H., Moggs, J., De Coen, W., Iguchi, T. & Callaghan, A. (2008). Linking molecular and population stress responses in *Daphnia magna* exposed to cadmium. *Environmental Science and Technology*, Vol. 42, No. 6, (March 2008), pp. 2181-2188, ISSN 0013-936X

Croteau, M.-E. & Luoma, S.N. (2009). Predicting dietborne metal toxicity from metal influxes. *Environmental Science and Technology*, Vol. 43, No. 13, (July 2009), pp. 4915-4921, ISSN 0013-936X

Darwin, C. (1859). *On the origin of species by means of natural selection*. John Murray, London

da Silva, E.M., Soares, A.M.V.M. & Moreno, A.J.M. (1998). The use of the mitochondrial transmembrane electric potential as an effective biosensor in ecotoxicological research. *Chemosphere*, Vol. 36, No. 10, (April 1998), pp. 2375-2390, ISSN 0045-6535

De Lange, H.J., Sperber, V. & Peeters, E. (2006). Avoidance of polycyclic aromatic hydrocarbon-contaminated sediments by the freshwater invertebrates *Gammarus pulex* and *Asellus aquaticus*. *Environmental Toxicology and Chemistry*, Vol. 25, No. 2, (February 2006), pp. 452-457, ISSN 0730-7268

Duan, Y.H., Guttman, S.I., Oris, J.T. & Bailer, A.J. (2001). Differential survivorship among allozyme genotypes of *Hyalella azteca* exposed to cadmium, zinc or low pH. *Aquatic Toxicology*, Vol. 54, No. 1-2, (September 2001), pp. 15-28, ISSN 0166-445X

Dunk, M.J., McMath, S.M. & Arikans, J. (2008). A new management approach for the remediation of polluted surface water outfalls to improve river water quality. *Water and Environment Journal*, Vol. 22, No. 1, (March 2008), pp. 32-41, ISSN 1747-6593

Gale, S.A., King, C.K. & Hyne, R.V. (2006). Chronic sublethal sediment toxicity testing using the estuarine amphipod, *Melita plumulosa* (Zeidler): evaluation using metal-spiked and field-contaminated sediments. *Environmental Toxicology and Chemistry*, Vol. 25, No. 7, (July 2006), pp. 1887-1898, ISSN 0730-7268

Garrett-Jones, C. & Gramiccia, G. (1954). Evidence of the development of resistance to DDT by *Anopheles sacharovi* in the Levant. *Bulletin of the World Health Organisation*, Vol. 11, No. 4-5, (1954), pp. 865- 883, ISSN 0042-9686

Hercus, M.J. & Hoffman, A.A. (2000). Maternal and grandmaternal age influence offspring fitness in *Drosophila*. *Proceedings of the Royal Society of London B*, Vol. 267, No. 1457 (October 2000), pp.2105-2110, ISSN 0962-8452.

Holt, E.A. & Miller, S.W. (2011) Bioindicators: Using organisms to measure environmental impacts. In: *Nature Education Knowledge*, 01-08-2011, Available from http://www.nature.com/scitable/knowledge/library/bioindicators-using-organisms-to-measure-environmental-impacts-16821310

Hutchinson, T.H., Solbé, J. & Kloepper-Sams, P.J. (1998). Analysis of the ecetoc aquatic toxicity (EAT) database III — Comparative toxicity of chemical substances to different life stages of aquatic organisms. *Chemosphere*, Vol. 36, No. 1, (January 1998), pp. 129-142, ISSN 0045-6535

Hyne, R.V., Gale, S.A. & King, C.K. (2005). Laboratory culture and life-cycle experiments with the benthic amphipod *Melita plumulosa* (Zeidler). *Environmental Toxicology and Chemistry*, Vol. 24, No. 8, (August 2005), pp. 2065-2073, ISSN 0730-7268

Iguchi, T., Watanabe, H. & Katsu, Y. (2007). Toxicogenomics and ecotoxicogenomics for studying endocrine disruption and basic biology. *General and Comparative Endocrinology*, Vol. 153, No. 1-3, (August-September 2007), pp. 25-29, ISSN 0016-6480

Jones, R. (2010). Environmental contamination associated with a marine landfill ('seafill') beside a coral reef. *Marine Pollution Bulletin*, Vol. 60, No.11, (November 2010), pp. 1993-2006, ISSN 0025-326X

King, C.K., Dowse, M.C., Simpson, S.L. & Jolley, D.F. (2004). An assessment of five Australian polychaetes and bivalves for use in whole-sediment toxicity tests: Toxicity and accumulation of copper and zinc from water and sediment. *Archives of Environmental Contamination and Toxicology*, Vol. 47, No. 3, (October 2004), pp. 314-323, ISSN 0090-4341

King, C.K., Simpson, S.L., Smith, S.V., Stauber, J.L. & Batley, G.E. (2005). Short-term accumulation of Cd and Cu from water, sediment and algae by the amphipod *Melita plumulosa* and the bivalve *Tellina deltoidalis*. *Marine Ecology Progress Series*, Vol. 287, (February 2005), pp. 177-188, ISSN 0171-8630

King, C.K., Gale, S.A., Hyne, R.V., Stauber, J.L., Simpson, S.L. & Hickey, C.W. (2006a). Sensitivities of Australian and New Zealand amphipods to copper and zinc in waters and metal-spiked sediments. *Chemosphere*, Vol. 63, No. 9, (June 2006), pp.1466-1476, ISSN 0045-6535

King, C.K., Gale, S.A. & Stauber, J.L. (2006b). Acute toxicity and bioaccumulation of aqueous and sediment-bound metals in the estuarine amphipod *Melita plumulosa*. *Environmental Toxicology*, Vol. 21, No. 5, (October 2006), pp. 489-504, ISSN 1522-7278

Kreitsberg, R., Zemit, I., Freiberg, R., Tambets, M. & Tuvikene, A. (2010). Responses of metabolic pathways to polycyclic aromatic compounds in flounder following oil spill in the Baltic Sea near the Estonian coast. *Aquatic Toxicology*, Vol. 99, No. 4 (September 2010), pp. 473-478, ISSN 0166-445X

Livadas, G.A. & Georgopoulos, G. (1953). Development of resistance to DDT by *Anopheles sacharovi* in Greece. *Bulletin of the World Health Organization*, Vol. 8, No. 4, (April 1953), pp. 497-511, ISSN 0042-9686

Lowry, J.K., Berents, P.B. & Springthorpe, R.T. (2000). Australian Amphipoda: Melitidae. In: *Crustacea.net*, 01-08-2011, Available from
http://www.crustacea.net/crustace/amphipoda/melitidae/index.htm

Lubchenco, J., McNutt, M., Lehr, B., Sogge, M., Miller, M., Hammond, S. & Conner, W. (2010). BP Deepwater Horizon oil budget: What happened to the oil? In *RestoreTheGulf.gov*, 15-08-2011, Available from
http://www.deepwaterhorizonresponse.com/posted/2931/Oil_Budget_descripti on_8_3_FINAL.844091.pdf.

Maes, G.E., Raeymaekers, J.A.M., Pampoulie, C., Seynaeve, A., Goemans, G., Belpaire, C. & Volckaert, F.A.M. (2005). The catadromous European eel *Anguilla anguilla* (L.) as a model for freshwater evolutionary ecotoxicology: Relationship between heavy metal bioaccumulation, condition and genetic variability. *Aquatic Toxicology*, Vol. 73, No. 1, (June 2005), pp. 99-114, ISSN 0166-445X

Mandaville, S.M. (June 2002). *Benthic macroinvertebrates in freshwaters – Taxa tolerance values, metrics and protocols*. Soil & Water Conservation Society of Metro Halifax, Retrieved from
http://www.chebucto.ns.ca/ccn/info/Science/SWCS/H-1/tolerance.pdf

Mann, R.M. & Hyne, R.V. (2008). Embryological development of the Australian amphipod *Melita plumulosa* Zeidler, 1989 (Amphipoda, Gammaridea, Melitidae). *Crustaceana*, Vol.81, No. 1 (January 2008), pp. 57-66, ISSN 0011-216X.

Mann, R.M., Hyne, R.V., Spadaro, D.A. & Simpson, S.L. (2009). Development and application of a rapid amphipod reproduction test for sediment-quality assessment. *Environmental Toxicology and Chemistry*, Vol. 28, No. 6, (June 2009), pp. 1244-1254, ISSN 1552-8618

Mann, R.M., Hyne, R.V., Simandjuntak, D.L. & Simpson, S.L. (2010). A rapid amphipod reproduction test for sediment quality assessment: In situ bioassays do not replicate laboratory bioassays. *Environmental Toxicology and Chemistry*, Vol. 29, No. 11, (November 2010), pp. 2566-2574, ISSN 1552-8618

Mann, R.M., Hyne, R.V. & Ascheri, L.M.E. (2011). Foraging, feeding, and reproduction on silica substrate: Increased waterborne zinc toxicity to the estuarine epibenthic amphipod *Melita plumulosa. Environmental Toxicology and Chemistry*, Vol. 30, No. 7, (July 2011), pp. 1649-1658, ISSN 1552-8618

Matson, C.W., Palatnikov, G., Islamzadeh, A., McDonald, T.J., Autenrieth, R.L., Donnelly, K.C. & Bickham, J.W. (2005). Chromosomal damage in two species of aquatic turtles (*Emys orbicularis* and *Mauremys caspica*) inhabiting contaminated sites in Azerbaijan. *Ecotoxicology*, Vol. 14, No. 5 (July 2005), pp. 1-13, ISSN 0963-9292

Matson, C.W., Lambert, M.M., McDonald, T.J., Autenreith, R.L., Donnelley, K.C., Islamzadeh, A., Politov, D.I. & Bickham, J.W. (2006). Evolutionary toxicology: population-level effects of chronic contaminant exposure on the marsh frogs (*Rana ridibunda*) of Azerbaijan. *Environmental Health Perspectives*, Vol. 114, No.4, (April 2006), pp 547-552, ISSN 0091-6765

Mhadhbi, L., Boumaiza, M. & Beiras, R. (2010). A standard ecotoxicological bioassay using early life stages of the marine fish *Psetta maxima. Aquatic Living Resources*, Vol. 23, No. 2, (April 2010), pp. 209-216, ISSN 0990-7440

Middleditch, B.S., Missler, S.R., Hines, H.B., McVey, J.P., Brown, A., Ward, D.G. & Lawrence, A.L. (1980). Metabolic profiles of penaeid shrimp: Dietary lipids and ovarian maturation. *Journal of Chromatography A*, Vol. 195, No.3 (July 1980), pp. 359–368, ISSN 0021-9673.

de Souza-Pinto, N.C., Eide, L., Hogue, B.A., Thybo, T., Stevnsner, T., Seeberg, E., Klungland, A. & Bohr, V.A. (2001). Repair of 8-Oxodeoxyguanosine Lesions in Mitochondrial DNA Depends on the Oxoguanine DNA Glycosylase (OGG1) Gene and 8-Oxoguanine Accumulates in the Mitochondrial DNA of OGG1-defective Mice. *Cancer Research*, Vol. 61, No. 14 (July 2001), pp. 5378-5381, ISSN 0891-5849.

Owens, R.W. & Dittman, D.E. (2003). Shifts in the diets of slimy sculpin (*Cottus cognatus*) and lake whitefish (*Coregonus clupeaformis*) in Lake Ontario following the collapse of the burrowing amphipod *Diporeia. Aquatic Ecosystem Health & Management*, Vol. 6, No. 3, (September 2003), pp. 311-323, ISSN 1463-4988

Phillips, N.R. & Hickey, C.W. (2010). Genotype-dependent recovery from acute exposure to heavy metal contamination in the freshwater clam *Sphaerium novaezelandiae*. *Aquatic Toxicology*, Vol. 99, No. 4, (September 2010), pp. 507-513, ISSN 0166-445X

Pichaud, N., Ballard, J.W.O., Tanguay, R.M. & Blier, P.U. (2011). Thermal sensitivity of mitochondrial functions in permeabilized muscle fibers from two populations of *Drosophila simulans* with divergent mitotypes. *American Journal of Physiology – Regulatory, Integrative and Comparative Physiology*, Vol. 301, No. 1 (July 2011), pp. R48-R59, ISSN 0363-6119.

Pillay, D., Branch, G.M., Griffiths, C.L., Willaims, C. & Prinsloo, A. (2010). Ecosystem change in a South African marine reserve (1960-2009): Role of seagrass loss and anthropogenic disturbance. *Marine Ecology Progress Series*, Vol. 415, (September 2010), pp. 35-48, ISSN 0171-8630

Pollino C.A. & Holdway, D.A (2002), Toxicity testing of crude oil and related compounds using early life stages of the crimson-spotted rainbowfish (*Melanotaenia fluviatilis*). *Ecotoxicolgy and Environmental Safety*, Vol. 52, No. 3, (July 2002), pp. 180-189, ISSN 0147-6513

Poynton, H.C., Loguinov, A.V., Varshavsky, J.R., Chan, S., Perkins, E.J. & Vulpe, C.D. (2008). Gene expression profiling in *Daphnia magna* Part I: Concentration-dependent profiles provide support for the no observed transcriptional effect level. *Environmental Science and Technology*, Vol. 42, No. 16, (August 2008), pp. 6250-6256, ISSN 0013-936X

Raj, A. & van Oudenaarden, A. (2008). Nature, nurture or chance: Stochastic gene expression and its consequences. *Cell*, Vol. 135, No. 2 (October 2008), pp. 216-226, ISSN 0092-8674.

Ringwood, A.H., Conners, D.E. & Hoguet, J. (1998). Effects of natural and anthropogenic stressors on lysosomal destabilization in oysters *Crassostrea virginica*. *Marine Ecology Progress Series*, Vol. 166, (May 1998), pp. 163-171, ISSN 0171-8630

Rinner, B.P., Matson, C.W., Islamzadeh, A., McDonald, T.J., Donnelly, K.C. & Bickham, J.W. (2011). Evolutionary toxicology: contaminant-induced genetic mutations in mosquitofish from Sumgayit, Azerbaijan. *Ecotoxicology*, Vol. 20, No. 2, (March 2011), pp. 365-376, ISSN 0963-9292

Robinson, J., Waller, M.J., Parham, P., Groot, N.d., Bontrop, R., Kennedy, L.J., Stoehr, P. & Marsh, S.G.E. (2003). IMGT/HLA and IMGT/MHC: Sequence databases for the study of the major histocompatibility complex. *Nucleic Acids Research*, Vol. 31, No. 1, (January 2003), pp. 311-314, ISSN 0305-1048

Roach, A.C., Jones, A.R. & Murray, A. (2001). Using benthic recruitment to assess the significance of contaminated sediments: the influence of taxonomic resolution. *Environmental Pollution*, Vol. 112, No. 2, (April 2001), pp. 131-143, ISSN 0269-7491

Roelofs, D., Janssens, T.K.S., Timmermans, M.J.T.N., Nota, B., Mariën, J., Bochdanovits, Z., Ylstra, B. & Van Straalen, N.M. (2009). Adaptive differences in gene expression associated with heavy metal tolerance in the soil arthropod *Orchesella cincta*. *Molecular Ecology*, Vol. 18, No. 15, (August 2009), pp. 3227-3239, ISSN 1365-294X

Rose, W.L. & Anderson, S.I. (2005). Genetic ecotoxicology, In: *Encyclopaedia of toxicology, 2nd edn*, P. Wexler, (Ed). 126-132, Elsevier Ltd, ISBN 0-12-745354-7, Oxford, United Kingdom

Schizas, N.V., Chandler, G.T., Coull, B.C., Klosterhaus, S.L. & Quattro, J.M. (2001). Differential survival of three mitochondrial lineages of a marine benthic copepod exposed to a pesticide mixture. *Environmental Science and Technology*, Vol. 35, No. 3, (February 2001), pp. 535-538, ISSN 0013-936X

Simpson, S.L., Batley, G.E., Chariton, A.A., Stauber, J.L., King, C.K., Chapman, J.C., Hyne, R.V., Gale, S.A., Roach, A.C. & Maher, W.A. (2005). *Handbook For Sediment Quality Assessment*. CSIRO, ISBN 0-643-09197-1, Bangor, Australia

Smith, K.E.C., Dom, N., Blust, R. & Mayer, P. (2010). Controlling and maintaining exposure of hydrophobic organic compounds in aquatic toxicity tests by passive dosing. *Aquatic Toxicology*, Vol. 98, No. 1, (June 2010), pp. 15-24, ISSN 0166-445X

Snape, J.R., Maund, S.J., Pickford, D.B. & Hutchinson, T.H. (2004). Ecotoxicogenomics: The challenge of integrating genomics into aquatic and terrestrial ecotoxicology. *Aquatic Toxicology*, Vol. 67, No. 2, (April 2004), pp. 143-154, ISSN 0166-445X

Snyder, C.D. & Hendricks, A.C. (1997). Genetic responses of *Isonychia bicolor* (Ephemeroptera:Isonychiidae) to chronic mercury pollution. *Journal of the North American Benthological Society*, Vol. 16, No. 3, (September 1997), pp. 651-663, ISSN 0887-3593

Spadaro, D.A., Micevska, T. & Simpson, S.L. (2008). Effect of nutrition on toxicity of contaminants to the epibenthic amphipod *Melita plumulosa*. *Archives of Environmental Contamination and Toxicology*, Vol. 55, No. 4, (November 2008), pp. 593-602, ISSN 0090-4341

Theodorakis, C.W., Blaylock, B.G. & Shugart, L.R. (1997). Genetic ecotoxicology I: DNA integrity and reproduction in mosquitofish exposed in situ to radionuclides. *Ecotoxicology*, Vol. 6, No., 4, (August 1997), pp. 205-218, ISSN 0963-9292

Theodorakis, C.W. & Shugart, L.R. (1997). Genetic ecotoxicology II: Population genetic structure in mosquitofish exposed *in situ* to radionuclides. *Ecotoxicology*, Vol. 6, No. 6, (December 1997), pp. 335-354, ISSN 0963-9292

Theodorakis, C.W., Bickham, J.W., Lamb, T., Medica, P.A. & Lyne, T.B. (2001). Integration of genotoxicity and population genetic analyses in kangaroo rats (*Dipodomys merriami*) exposed to radionuclide contamination at the Nevada Test Site, USA. *Environmental Toxicology and Chemistry*, Vol. 20, No. 2, (February 2001), pp.317-326, ISSN 1552-8618

Theodorakis, C.W. & Wirgin, I. (2002). Genetic responses as population-level biomarkers of stress in aquatic ecosystems. In: *Biological Indicators of Aquatic Ecosystem Health*, S.M. Adams, (Ed). 147-186, American Fisheries Society, ISBN 1-888569-43-3, New York, United States of America

Toro, B., Navarro, J.M. & Palma-Fleming, H. (2003). Relationship between bioenergetics responses and organic pollutants in the giant mussel, *Choromytilus chorus* (Mollusca: Mytilidae). *Aquatic Toxicology*, Vol. 63, No. 3, (May 2003), pp. 257-269, ISSN 0166-445X

Van Aggelen, G., Ankley, G.T., Baldwin, W.S., Bearden, D.W., Benson, W.H., Chipman, J.K., Collette, T.W., Craft, J.A., Denslow, N.D., Embry, M.R., Falciani, F., George, S.G., Helbing, C.C., Hoekstra, P.F., Iguchi, T., Kagami, Y., Katsiadaki, I., Kille, P., Liu, L., Lord, P.G., McIntyre, T., O'Neill, A., Osachoff, H., Perkins, E.J., Santos, E.M., Skirrow, R.C., Snape, J.R., Tyler, C.R., Versteeg, D., Viant, M.R., Volz, D.C., Williams, T.D. & Yu, L. (2009). Integrating omic technologies into aquatic ecological

It's All in the Genes: How Genotype Can Impact Upon Response to Contaminant Exposure and the Implications for
Biomonitoring in Aquatic Systems

155

risk assessment and environmental monitoring: Hurdles, achievements, and future outlook. *Environmental Health Perspectives*, Vol. 118, No.1, (January 2010), pp. 1-5, ISSN 0091-6765

Vandegehuchte, M.B., De Coninck, D., Vandenbrouck, T., De Coen, W.M. & Janssen, C.R. (2010). Gene transcription profiles, global DNA methylation and potential transgenerational epigenetic effects related to Zn exposure history in *Daphnia magna*. *Environmental Pollution*, Vol. 158, No. 10 (October 2010), pp. 3323-3329, ISSN 0269-7491

van Straalen, N.M. & Timmermans, M.J.T.N. (2002). Genetic variation in toxicant-stressed populations: An evaluation of the "genetic erosion" hypothesis. *Human and Ecological Risk Assessment: An International Journal*, Vol. 8, No. 5, (July 2002), pp. 983–1002, ISSN 1080-7039

Vijayavel, K., Gopalakrishnan, S. & Balasubramanian, M.P. (2007). Sublethal effect of silver and chromium in the green mussel *Perna viridis* with reference to alterations in oxygen uptake, filtration rate and membrane bound ATPase system as biomarkers. *Chemosphere*, Vol. 69, No. 6, (October 2007), pp. 979-986, ISSN 0045-6535

Wallace, J.B. & Webster, J.R. (1996). The role of macroinvertebrates in stream ecosystem function. *Annual Review of Entomology*, Vol. 41, (January 1996), pp. 115-139, ISSN 0066-4170

Watanabe, H., Kobayashi, K., Kato, Y., Oda, S., Abe, R., Tatarazako, N. & Iguchi, T. (2008). Transcriptome profiling in crustaceans as a tool for ecotoxicogenomics. *Daphnia magna* DNA microarray. *Cell Biology and Toxicology*, Vol. 24, No. 6, (December 2008), pp. 641-647, ISSN 0742-2091

Wilkie, E.M., Roach, A.C., Micevska, T., Kelaher, B.P. & Bishop, M.J. (2010). Effects of a chelating resin on metal bioavailability and toxicity to estuarine invertebrates: Divergent results of field and laboratory tests. *Environmental Pollution*, Vol. 158, No. 5, (May 2010), pp. 1261-1269, ISSN 0269-7491

Williams, R., Gero, S., Bejder, L., Calambokidis, J., Kraus, S.D., Lusseau, D., Read, A.J. & Robbins, J. (2011). Underestimating the damage: interpreting cetacean carcass recoveries in the context of the Deepwater Horizon/BP incident. *Conservation Letters*, Vol. 4, No. 3, (June-July 2011), pp. 228-233, ISSN 1755-263X

Wirgin, I. & Waldman, J.R. (2004). Resistance to contaminants in North American fish populations. *Mutation Research*, Vol. 552, No.1-2, (August 2004), pp. 73-100, ISSN 0027-5107.

Wu, R.S.S. & Or, Y.Y. (2005). Bioenergetics, growth and reproduction of amphipods are affected by moderately low oxygen regimes. *Marine Ecology Progress Series*, Vol. 297, (August 2005), pp. 215-223, ISSN 0171-8630

Yauk, C., Polyzos, A., Rowan-Carroll, A., Somers, C.M., Godschalk, R.W., Van Schooten, F.J., Berndt, M.L., Pogribny, I.P., Koturbash, I., Williams, A., Douglas, G.R. & Kovalchuk, O. (2008). Germ-line mutations, DNA damage, and global hypermethylation in mice exposed to particulate air pollution in an urban/industrial location. *Proceedings of the National Academy of Sciences*, Vol. 105, No. 2, (January 2008), pp. 605-610, ISSN 0027-8424

Zeidler, W. (1989). A new species of *Melita* (Crustacea: Amphipoda: Melitidae) from northern New South Wales with a note on the genus *Abludomelita* Karaman, 1981.

Proceedings of the Linnean Society of New South Wales, Vol. 110, No. 4, pp. 327-338, ISSN 0370-047X

Zhou, J., Cai, Z.-H., Li, L., Gao, Y.-F. & Hutchinson, T.H. (2010). A proteomics based approach to assessing the toxicity of bisphenol A and diallyl phthalate to the abalone (*Haliotis diversicolor supertexta*). *Chemosphere*, Vol. 79, No.5, (April 2010), pp. 595-604, ISSN 0045-6535

Part 4

Microbial Contamination

9

Cryptosporidium spp. and *Giardia duodenalis*: A picture in Portugal

André Silva Almeida[1,2], Sónia Cristina Soares[2],
Maria Lurdes Delgado[1], Elisabete Magalhães Silva[2],
António Oliveira Castro[1] and José Manuel Correia da Costa[1]
[1]*Center for Parasite Immunology and Biology, CSPGF-INSA, Porto*
[2]*Center for the Study of Animal Science, CECA-ICETA, Porto*
Portugal

1. Introduction

Cryptosporidium is an entero-pathogen which causes gastrointestinal disturbs. Primarily this organism infects the microvillous border of the intestinal epithelium, and to lesser extent extra intestinal epithelia, causing acute gastrointestinal disturbs (Fayer, 2004). The duration of infection and the ultimate outcome of intestinal cryptosporidiosis greatly depend on the immune status of the patient. In fact, immunologically healthy patients usually recover spontaneously in a week. The clinical signs can range from asymptomatic to acute, severe and persistent diarrhea and their potential for *Cryptosporidium* transmission can persist for weeks after symptoms cease (Deng et al., 2004; Fayer, 2004; Hunter and Thompson, 2005).

Similar to *Cryptosporidium*, *Giardia* is an entero-pathogen, but a non-cell-invasive which causes giardiasis. The most prominent clinical signs of the disease are abdominal pain, nausea, followed by severe watery diarrhea, dehydration, malabsorption (particularly lipids and lipid soluble vitamins) and weight loss. Chronic courses are characterized by recurrent brief or persistent episodes of diarrhoea (Eckmann and Gillin, 2001; Muller and von Allmen, 2005).

Due to some particular biological features, *Cryptosporidium* and *Giardia* represent an important question of environmental contamination with impact for human and animal health. These are single-celled organisms that belong to the kingdom Protista, and are common food and waterborne protozoa that affect humans and a wide range of domestic and wild animals (Fayer, 2004). They have a low infection dose necessary to infect humans, with possible as few as 10 organisms in some cases (USDA). They have emerged in the last two decades as intriguing microbes with an enormous impact in Animal (including wildlife species) and Human Health, potentiated by the environmental contamination of the infectious stages. Both can cause mild to severe diarrhea. No specific therapy has proven to be effective, but immunocompetent individuals generally recover within a week (USDA). However, immunocompromised individuals may be unable to clear the parasites and, therefore, suffer chronic and debilitating illness. They have been recognized as important pathogens in contaminated drinking water, meaning a problem of environmental contamination, due to two main reasons pointed out: 1) their resistance and biological viability under conventional drinking water treatment conditions (chlorination and filtration); 2) the occurrence of cryptosporidiosis and giardiasis outbreaks associated with the consumption of contaminated water. This was the case in an outbreak in Milwaukee (Wisconsin, USA) in 1993, the largest waterborne disease outbreak reported all over the World. An estimated 400,000 people were reported ill (USDA).

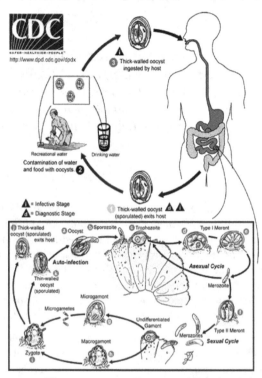

Fig. 1. Life cycle of *Cryptosporidium parvum* or *Cryptosporidium hominis* . From CDC (www.cdc.gov).

The life cycle of *Cryptosporidium* comprises an asexual stage and sexual stage; *Cryptosporidium* has a spore phase named oocyst, which represents the infectious stage of *Cryptosporidium* and is the resistant form found in the environmental; the oocyst is extraordinary resistant to common disinfectants of water, such as chloride, and the water represents the primary transmission route (Figure 1). The infection begins with the ingestion of oocysts through contaminated water or food, or by fecal-oral contact. After ingestion, the excystation of oocyst is induced by the acidic nature of stomachic lumen and the presence of enzymes. However, this event occurs in the small intestine favored by the presence of the neutral pH, bile salts and fatty acids. Each oocyst contains four sporozoites. They are released and try immediately to infect epithelial cells of the gastrointestinal tract. At the end of this endogenous cycle, sporulated oocysts are formed which, once shed in the environment with feces, are ready to infect a new suitable host. The prepatent period, which means the time between the ingestion of infecting oocysts and the excretion of a new generation of oocysts, varies with the host and species of *Cryptosporidium*, but usually it ranges from 4 to 22 days. The patent period, which means the duration of oocyst excretion, ranges from 1 to 20 days. As previously stated, transmission of *Cryptosporidium* may occur through contact with contaminated water and food. In fact, many reported outbreaks occurred in water parks, pools and day care centers.

The life cycle of *Giardia* is monoxenic, and comprises two stages: the cyst and the trophozoite (Figure 2). The cysts are the resistant form and are approximately 7 to 10 μm length and oval in shape; cysts can be found in feces and are released into the environment where they can survive and remain viable for several months in cool or moist conditions. They are responsible for disease transmission, and they are able to survive under the standard concentrations of chlorine used in water treatment plants. The infection occurs after ingestion of cysts in contaminated water, food, or by the fecal-oral route in the absence of hygienic conditions. The excystation of cysts is induced by the acidic nature of stomachic lumen and the presence of enzymes. However, this event occurs in the small intestine favored by the presence of the neutral pH, and bile salts and fatty acids. Each cyst contains two trophozoites. Trophozoites have two distinct nuclei, four pairs of flagellae, are 12 to 15 μm length. They multiply by asexual reproduction, longitudinal binary fission, and colonize the lumen of the proximal small bowel, attaching to the mucosa of the bowel using a ventral sucking disk. Trophozoites are responsible for the clinical disease in the host. Cysts are excreted in the feces and became immediately infectious, making possible the transmission from person-to-person (http://www.dpd.cdc.gov/dpdx/HTML/Giardiasis.htm). The prepatent period varies with the host and species of *Giardia*, with a median value of 14 days.

Giardia and *Cryptosporidium* share common characteristics that influence greatly the epidemiology of their infections. They are maintained in a variety of transmission cycles, independently, not requiring interaction between them (Figure 3). *Giardia* can be maintained in independent cycles involving wildlife or domestic animals, and similarly, *Cryptosporidium* can be maintained in cycles involving livestock, especially cattle. As observed in Figure 3, the circumstances under which such cycles may interact and where zoonotic transfer occurs are not completely understood (Hunter and Thompson, 2005). Cysts and oocysts are the stage transmitted from an infected host to a susceptible host by the fecal-oral route. Several common transmission routes exist, and include a) person-to-person through direct or indirect contact, where sexual activities may potentiate transmission, b) from animal-to-animal, c) animal-to-human, d) water-borne through drinking water and recreational water, and, e) food-borne (Caccio and Ryan, 2008; Caccio et al., 2005; Fayer et al., 2000; Hunter and Thompson, 2005). The infected hosts, whether human or animals, shed very large numbers of transmissive stages (oocysts and cysts) in their faeces, thereby increasing environmental contamination. The infective dose of both parasites, in human infections, was calculated taking into account statistical data and experimental infection studies: the ID50 varies regarding the isolates, ranging from 9 to 1042 oocysts for *Cryptosporidium* and 1 to 10 cysts for *Giardia* (Adam, 2001; Fayer et al., 2000; Okhuysen et al., 1999). These features markedly influence the epidemiology of these infections: a) the infective dose is low for both parasites; b) cysts and oocysts are immediately infectious when excreted in faeces, and possess several transmission routes; c) cysts and oocysts are very stable and can survive for weeks to months in the environment; d) water and food may became contaminated due to the environmental dispersal. The transmission of these infections, either direct or indirect, is favored by several factors such as high population densities and close contact with infected hosts or contaminated water or food. These factors are dependent on the infecting species, either in zoonotic and anthroponotic transmissions. Recent studies suggested separated risks for *C. hominis* (such as travel abroad and contact with infected diarrheic individuals) and *C. parvum* (contact with cattle) (Caccio et al., 2005; Hunter et al., 2004). In sporadic cryptosporidiosis, risk factors include the age of patients (children under five years of age), travelling, contact with infected individuals and contact with farm animals (Caccio et al., 2005). Furthermore, swimming in public swimming pools or recreational areas represents a risk of infection, as suggested by Australian and US studies (Robertson et al., 2002; Roy et al., 2004). Curiously, authors have postulated that, although *Cryptosporidium* is transmitted through contaminated food, a small number of parasites in these samples may not induce infection with clinical symptoms but a protective immunity (Meinhardt et al., 1996).

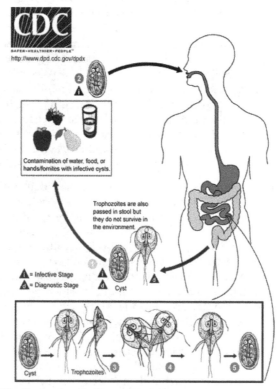

Fig. 2. Life cycle of *Giardia duodenalis*. From CDC (www.cdc.gov).

There are no requirements for testing surface waters for the presence of these parasites, although analyses of outbreaks have shown that the pathogens can be shed into recreational waters (Castro-Hermida et al., 2008). Also, similar studies regarding *Giardia* transmission and sporadic giardiasis performed by authors in UK, reveled as main risks the swallowing of water while swimming, drinking treated tap water, contact with fresh water and easting lettuce (Stuart et al., 2003). The introduction of molecular tools analysis on the epidemiological field can produce useful information allowing to a better understanding about the origin of contamination, the genetic characterization of involved species/genotype/assemblage and their zoonotic potential: a new field in the modern molecular epidemiology.

The taxonomic and filogenetic relationships of *Cryptosporidium* and *Giardia* remain poorly defined; thus, the understanding of their transmission dynamics has been limited. A consensus has been adopted: with molecular techniques, the ability to observe extensive genetic variation within *Cryptosporidium* and *Giardia* species is leading to a better understanding of the taxonomy and zoonotic potential of these variants, and the epidemiology of the diseases. Namely, genotyping of samples using molecular analysis at informative loci is necessary to distinguish species and genotypes that are involved and their zoonotic potential.

These molecular tools have been helpful to enhance our knowledge and understanding of the taxonomy, host range and transmission routes of *Cryptosporidium* and *Giardia* and the epidemiology of human disease. Moreover, these tools are used to understand the public health importance of different environmental routes of transmission, leading toward improved strategies for prevention and surveillance of cryptosporidiosis and giardiasis (Fayer et al., 2000; Jex et al., 2008; Monis and Thompson, 2003; Smith et al., 2006). Some methods rely on the specific *in situ* hybridization of probes to particular genetic loci within *Cryptosporidium* oocysts and *Giardia* cysts, whereas the majority relies on the specific amplification of one or more loci from small amounts of genomic DNA by polymerase chain reaction (PCR). This is particularly helpful in the environmental samples or others where the parasite load is low (Jex et al., 2008; Smith et al., 2006).

The present chapter presents a long term study performed in Portugal aiming to assess the prevalence of two Protozoan parasites, *Cryptosporidium* spp. and *Giardia duodenalis*, in surface raw water and drinking water samples, and assess the risk that this microbial environmental contamination poses for human and animal health. The study was guided by a multidisciplinary team comprising several Veterinarians, researchers and technicians, and developed in the North region of Portugal. A long term program was established aiming at pinpointing the sources of surface water, drinking water, and environmental contamination, working with the water-supply industry. This program comprised the collection of raw and treated water samples in previously defined areas by using the Method 1623, described by the Environmental Protection Agency (US – EPA).

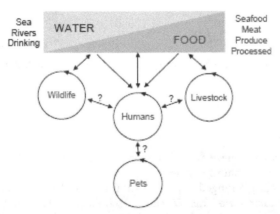

Fig. 3. The most important cycles of transmission for maintaining Giardia and Cryptosporidium. Besides direct transmission, water and food may also play a role in transmission. The frequency of interaction between cycles is not known. From (Hunter and Thompson, 2005)

2. Material and methods

In the north of Portugal there are 5 major hydrographical basins forming the most important water resources of the country. These hydrographical basins are named after the main rivers; Minho, Lima, Cávado, Ave, and Douro (Figure 4, 5). Cávado and Ave rivers run entirely inside national (Portuguese) borders, while Minho, Lima, and Douro are international rivers, with sources in Spain. Global features of the area, based on data provided by CCDRN, are as follows: The resident population in 2006 is 3,744,341 inhabitants; animal husbandry is an important economic activity; there are poorly developed sanitation infrastructures with wastewater plants discharging into these hydrographical basins; 64% of resident population has access to sanitation and 84% has access to treated drinking water; the region has the highest population density of the country, reaching 176 inhabitants per km², there are a high number of recreational areas, river beaches as well as pumping areas for drinking water plants.

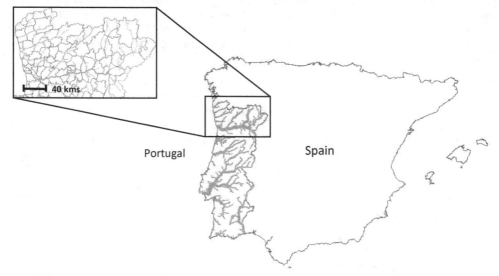

Fig. 4. Geographic location of the North of Portugal and its 5 hydrographical basins into Iberian Peninsula

Water sample collection

Raw water collection

Raw water samples were collected twice a year between January 2004 to December 2006 from 97 sources, including main rivers and respective affluents, from upriver to downriver. The volume of each sample ranged from 25 to 100 L. Samples were filtered through Filta-Max filters (IDEXX Laboratories, Inc., Westbrook, Maine, USA) with a pump on the inlet side of the filter according to the recommendation of the manufacturer. Intact filters were kept in refrigerated containers and transported immediately to the laboratory. The filter was taken from the container and processed with the aid of a Filta-Max Manual Wash Station (IDEXX Laboratories, Inc.) for further elution and concentration process, which consisted of decompression of the filter, passing the sample through a membrane, and centrifugation. A sample pellet (around 2 ml) was obtained and transferred to a Leighton tube for subsequent immunomagnetic separation (IMS).

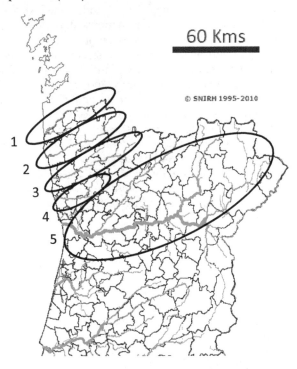

Fig. 5. Location of the 5 hydrographical basins in the North of Portugal. 1. Minho; 2. Lima; 3. Cávado; 4. Ave; 5. Douro

Drinking water collection

As previously described for raw water, drinking water samples were collected twice a year from January 2004 to December 2006 in 43 sampling points in the drinking water treatment plants, from upriver to downriver, from all the 5 hydrographical basins (Figure 4 and 5). Always as possible, drinking water sampling points were derived from the previous related raw water collection points. This was not always possible due to legal impositions from the water supply companies. The volume of each sample ranged from 80 to 100 L. Samples were collected locally, filtered through Filta-Max filters (IDEXX Laboratories, Inc., Westbrook, ME, USA) as previously described for raw water processing.

Parasite detection

The IMS procedure was performed according to the USEPA method 1623 [USEPA]. Briefly, anti-*Giardia* and anti-*Cryptosporidium* magnetic beads were mixed with SL Buffer A and SL Buffer B in each Leighton tube containing the sample concentrate (Dynabeads GC-Combo, Invitrogen Dynal, A. S., Oslo, Norway) and incubated 1 hr at room temperature. Then, using 2 magnetic particle concentrators, beads were collected, washed, and transferred into a 1.5

ml tube. Fifty-µl of 0.1 N HCl were added to each sample to dissociate beads from the target organisms, the beads were rejected, and the suspension was transferred to the wells of the slides containing 5 µl of 1.0 N NaOH. The samples were air-dried overnight and stained with FITC-conjugated anti-*Cryptosporidium* spp. and anti-*Giardia* spp. monoclonal antibodies, according the manufacturer's instructions (Crypto/*Giardia* Cells, Cellabs, Australia). The excess of FITC-mAb was removed by adding 100 µl of PBS to each well, leaving the slides for 5 min, and aspirating the excess of PBS. A 50-µl aliquot of 4'-6'-diamino-2-phenylindole (DAPI) solution (0.4 mg/ml in PBS) was introduced into each well. The slides were left at room temperature for 15 min, and excess DAPI solution removed by washing the slides in PBS. Slides were examined by epifluorescence microscope. *Giardia* cysts and *Cryptosporidium* oocysts were indentified and counted based on their shape and size using a Nikon Optiphot fluorescence microscope (Nikon Corporation, Tokyo, Japan). The number of cysts and oocysts per each well was recorded and concentrations extrapolated per 10 L of sample. Positive and negative controls were performed as indicated by the manufacturer and recommended in the Method 1623. The mean recovery percentages of oocysts of *Cryptosporidium* spp. and cysts of *Giardia* spp. using the Filta-Max system and IMS procedures from water samples is, according to the manufacturer, 50 ± 13% and 41 ± 79%, respectively (McCuin and Clancy, 2003).

DNA extraction, PCR and sequencing

PCR analysis was performed in the samples with the highest density of infectious stages of both parasites detected by DFA. The criterion utilized was the detection of a minimum of 50 cysts/oocysts of any parasite in the total sample volume. In this context, the genetic characterization was executed in 80 samples derived from raw water and 4 samples derived from treated drinking water. The cover slip was separated from the slide and with the aid of cotton swab soaked with 100 µl of distillated water, the surface of the slide was scraped in order to collect the sample. It was confirmed, under microscope observation that the slide had no remaining cysts or oocysts. The tip of the cotton swab was cut and placed in a 1.5 ml tube for subsequent DNA extraction with a QIAamp DNA Mini Kit (QIAGEN GmbH, Hilden, Germany), according to the manufacturer's instructions.

For determining the species of *Cryptosporidium* and *Giardia* present in the samples, PCR analyses were performed. A 2-step nested PCR was performed to amplify a portion of the small subunit (SSU) ribosomal RNA gene of *Cryptosporidium* spp. (Xiao et al., 1999). For the molecular typing of *Giardia* spp., a semi-nested PCR was performed to amplify a portion of the b-giardin gene (Caccio et al., 2002). For all PCR reactions, negative and positive controls were prepared, with sterile water and reference DNA, respectively. The PCR products were analyzed in agarose gel (1.4%) stained with ethidium bromide under UV light. Images were captured with a gel documentation system (GelDoc2000, BioRad, Hercules, California, USA). The PCR products of the successful reaction were purified by Wizard SV Gel and PCR Clean-up System (Promega, Madison, Wisconsin, USA) and sequenced in both strands by an external laboratory (EUROFINS MWG OPERON, Ebersberg, Germany). Chromatograms were examined with the software ChromasPro (http://www.technelysium.com.au/ChromasPro.html) and the sequences with the software ProSeq (http://www.biology.ed.ac.uk/research/institutes/evolution/software/filatov/ proseq.htm). Sequences were compared with the GenBank database with the tool BLAST (http://blast.ncbi.nlm.nih.gov/Blast.cgi) and deposited in the database Zoop-Net of the Med-Vet-Net network (http://www.medvetnet.org).

3. Results

IMS and DFA detection of infectious stages of *Cryptosporidium* spp. and *G. duodenalis* in raw and drinking water samples

The number of validated raw water samples in this study was 283. Environmental stages of the protozoa were detected in 72.8% (206 of 283) of the water samples, including 15.2% (43 of 283) positive for cysts of *G. duodenalis*, 9.5% (27 of 283) for oocysts of *Cryptosporidium* spp., and 48.1% (136 of 283) for both parasites (Table 1). In Figure 6, the percentages of positive and negative samples from the 5 hydrographical basins are shown individually. The Ave basin showed the highest percentage of positive samples; 90.2% of the samples were positive. Minho basin showed the lowest percentage of positive samples, even though this value was more than 64%. In all the 5 hydrographic basins, the co-presence of *Cryptosporidium* and *Giardia* counted for the majority of positive samples, with the exception of Minho basin in which *Giardia* positive samples were slightly more than *Cryptosporidium* and *Giardia* positive samples. In the cases where both parasites were present in the same sample, the number of *G. duodenalis* cysts always outnumbered *Cryptosporidium* spp. oocysts. We also found no correlation between the concentrations of both parasites, meaning that, when the concentration of *Giardia* cysts is high, *Cryptosporidium* oocysts are not necessarily high (data not shown). Furthermore, the range of the concentrations of *G. duodenalis* cysts was much higher than *Cryptosporidium* spp. oocysts (0.17-50,000 cysts per 10 L and 0.2-726.1 oocysts per 10 L, respectively). In positive water samples, no empty (i.e., without internal characteristics, or ghosts) or DAPI negative *Cryptosporidium* spp. oocysts or *G. duodenalis* cysts were found. In all cases, it was possible to observe an increase of parasite load from upriver to downriver. The majority of water samples from the international rivers (Minho, Lima, and Douro) collected at the border with Spain was negative.

	Positive	%	Cryptosporidium	%	Giardia	%	Cryptosporidium and Giardia	%	Total	%
Raw water	206	72,8	27	9,5	43	15,2	136	48,1	283	100,0
Drinking water	43	25,7	17	10,2	14	8,4	12	7,2	167	100

Table 1. Overall prevalence of infectious stages of *Cryptosporidium* spp. and *Giardia duodenalis* in raw and drinking water samples collected.

Similar to this, the number of validated drinking water samples in this study was 167. In Figure 6 the percentage and concentration of isolates from the 5 hydrographical basins are shown. Negative results, for both protozoa, were obtained in 124 out of 167 drinking water samples. Infectious stages of the protozoa were detected in 25.7% (43 out of 167) of the water samples. Among them, 8.4% (14 out of 167) of the samples were with cysts of *Giardia*, 10.2% (17 out of 167) with oocysts of *Cryptosporidium*, and 7.2% (12 out of 167) with both parasites (Table 1). In positive water samples, no empty (i.e., without internal characteristics, or ghosts) or DAPI negative oocysts of *Cryptosporidium* spp. or cysts of *G. duodenalis* were found. Furthermore, the mean concentrations of *G. duodenalis* cysts were much higher than those of the *Cryptosporidium* spp. oocysts (0.1-108.3 cysts per 10 L and 0.1-12.7 oocysts per 10 L, respectively).

Sample	Sampling site	Number of oocysts (C) and cysts (G) detected	Giardia duodenalis assemblages	Cryptosporidium species
51 B	Ponte do Bico R. Homem	398G - 334C		C. andersoni
53 B	Ribeira de Panoias	25000G - 35C	A - I & B	C. andersoni
55 B	Ribeira de pontes Barcelos	511G - 1955C		C.andersoni
56 B	Jusante Etar de Barcelos	11,000G - 96C		C. andersoni
58 B	Foz do Cávado-Barca do Lago	5754G - 98C		C. andersoni
60 B	Eta Areias de vilar r. Cávado	2313G - 32C	A-II	C. hominis
66 B	Rio Selho Pte Brandão	6590G - 90C	A-II	
67 B	Foz do Vizela/Rebordões	2178G - 46C	B	C. andersoni
68 B	Ponte da Trofa/R Ave	2146G - 1211C	A & B & E	C. andersoni
69 B	Ponte do Ave/R Ave	1022G - 84C	A & B & E	C. andersoni
70 B	Foz do R Este/Arcos/Vila Conde	855G - 262C	A-II	C. andersoni
79 B	Eta de Ferreira	5396G - 44C	A-II	C. andersoni
80 B	Albufeira Alto Cávado	221G - 37C		C. parvum
98 B	Foz do Sabor/Cabeça boa	233G - 16C		C. parvum
105 B	Foz do Sousa-Gondomar	1192G - 33C		C. andersoni
132 B	Ponte do Bico R. Homem-Bruta	526G - 11C		C. andersoni
139 B	Jusante Etar de Barcelos	1101G - 4C		C. andersoni
141 B	Foz do Cávado-Barca do Lago	686G - 13C		C. andersoni
147 B	Ponte do Ave/R Ave	1,131G - 56C		C. andersoni
148 B	Foz do R Este/Arcos/Vila Conde	449G - 494C		C. andersoni
154 B	Ponte do Brandão-Rib do Selho	5128G - 229C		C. muris

Table 2. Results of genetic characterization of *Cryptosporidium* spp. and *Giardia duodenalis* and the respective number of cysts and oocysts (same table was published in Almeida et al. Korean J Parasitol 2010; 48:91-95).

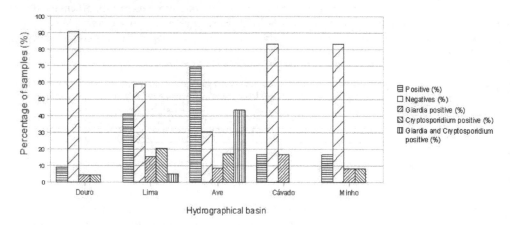

Fig. 6. Distribution of the results obtained by Method 1623 EPA-USA for infectious stages of *Cryptosporidium* spp. and *Giardia duodenalis* in drinking water samples collected in the 5 hydrographical basins (same figure was published in Almeida et al. Korean J Parasitol 2010; 48:165-169). It is shown the percentage of positive or negative samples per hydrograpical basin.

Genetic characterization of species and genotypes isolated

By PCR, it was unable to amplify DNA extracted from slides containing less than 50 oocysts of *Cryptosporidium* and 50 cysts of *Giardia*. Furthermore, positive amplifications over 3 replicates were never obtained when the number of cysts and oocysts was less than 1,000 per slide. With this criterion, of all the positive IMS samples from raw and drinking water, PCR amplification was performed over 80 samples and 4 samples derived from drinking water. Of referring that in these drinking water samples, the PCR was not successful and no amplification was obtained. In raw water samples, genetic characterization was successful in 8 samples for *G. duodenalis* and 20 samples for *Cryptosporidium* spp. In 59 samples, PCR amplification was not successful. A summary of the PCR results is shown in Table 2. *Cryptosporidium andersoni* was found in 16 samples, *Cryptosporidium parvum* in 2 samples, *Cryptosporidium hominis* in 1 sample, and *Cryptosporidium muris* in 1 sample. *G. duodenalis* assemblage A-II was found in 4 samples, assemblage B in 1 sample, and in the remaining 3 samples assemblages A, B, and E were found. Although it was not possible to obtain positive PCR results in treated water, it is expected that at the end of the treatment process the same species and genotypes will be found.

4. Discussion

The results of the present study indicate that the infectious stages of *Cryptosporidium* spp. and *G. duodenalis* are widely distributed in the rivers of northern Portugal in very significant concentrations. Also, it is observed that drinking water for human consumption is produced in very efficient plants regarding the elimination of these parasites.

The region has a high density of livestock farms favoring the cycle of parasite amplification. Surface waters (rivers, reservoirs, canals, and low land reservoirs) are used to produce drinking water for human consumption, and some of the raw water collection points presented in this work are the same collection points of the water industry for drinking water production. The surface water collected from the rivers is used as drinking water for the animals or used for agricultural purposes, by the majority of farmers, the feces are directly released into the rivers or reach it by runoff waters. The environmental contamination in these conditions is greater, as the contamination of husbandry. In fact, has been shown in a previous study a prevalence of 25% of *Cryptosporidium* infections in 467 bovine fecal samples in northern Portugal (Mendonca et al., 2007). Producers must be advertized about the risk of infection for both protozoa, and the low efficiency in animal production. It is very likely that these are the reasons why it was observed a considerable prevalence of these parasites in surface water, although the drinking water for human consumption derived from it has a low prevalence of these parasites: it is produced in very efficient plants of water industry. In the great majority of raw water samples it was observed contamination by *Cryptosporidium* or *Giardia* (72,8 %), but in the drinking water this number was quite low (25,7 %). Furthermore, in the positive samples, the parasite load was largely lower than the infectious dose.

Systematic studies on the genetic characterization of both protozoans indicate that the genus *Cryptosporidium* includes species that are infective for humans only (anthroponotic), humans and animals (zoonotic), and other pathogenic species that are not infective for humans (Fayer). Also, several *G. duodenalis* genotypes are infective to humans (zoonotic genotypes), and others are non-pathogenic (Caccio and Ryan, 2008). Amplification and sequencing of genes 18S SSU rRNA for *Cryptosporidium* spp. and b-giardin for *G. duodenalis* has been used to identify the zoonotic species and genotypes of the parasites (Hunter and Thompson, 2005; Read et al., 2004). Obviously, it is largely recognized that there is lack of consensus about genetic markers for the correct assignment of the species and subspecies of *Cryptosporidium* and *Giardia*. The gene markers (18S SSU rRNA and b-giardin) are generally accepted as good markers mainly because they are multicopy genes (18S SSU rRNA), restricted to these parasites (b-giardin), and with fixed differences among *Cryptosporidium* and *Giardia* species and subspecies (both genes). In the present study, in an attempt to produce relevant and comparable results, the choice of 18S SSU rRNA and b-giardin genes, frequently used by the most recognized researchers, was considered.

C. parvum (zoonotic) and *C. hominis* (anthroponotic) are the most common human-infecting species reported in river water samples in Europe, and *C. andersoni* is the animal-infecting species (Castro-Hermida et al., 2008; Xiao and Fayer, 2008). The same studies indicated G. *duodenalis* assemblage A as the most common zoonotic genotype, and *G. duodenalis* assemblage E as the most common non-zoonotic genotype (Almeida et al., 2006a; Caccio et al., 2003; Castro-Hermida et al., 2008; Sousa et al., 2006; Xiao and Fayer, 2008).

Our results suggest that the contamination of the surface waters in the north of Portugal is highly significant. We have found the zoonotic species of the genus *Cryptosporidium* described by other authors (Lobo et al., 2009). In addition, there was higher concentration of *G. duodenalis* detected by genotyping, with a greater genetic diversity. Assemblage A was found in 7 PCR positive samples (1 A-I and 6 A-II). The presence of the assemblage A-I has been suggested as an indicator of water contamination by livestock, while assemblage A-II has been considered a potential indicator of water contamination by humans (Almeida et al., 2006a; Caccio and Ryan, 2008). Nevertheless, in the northern part of the country assemblage A-I have been found in human samples (Sousa et al., 2006), and assemblage A-II in bovine samples (Castro-Hermida et al., 2007; Mendonca et al., 2007). Also, *G. duodenalis* assemblage B was detected in 4 of 8 samples. Assemblage B has been reported as a zoonotic genotype. The presence of infectious stages of this genotype in water samples has been attributed to water contamination by humans (Caccio and Ryan, 2008). Assemblage E was detected in 2 samples associated with assemblage A and B, suggesting a mixed human and animal source of contamination.

C. parvum was detected in 2 of the 20 processed samples. This species has a great zoonotic potential, and may have an animal or human source of contamination. A few studies concerning the biological reservoir (human and bovine) in the north of Portugal have indicated *C. parvum* as an important pathogen infecting the great majority of bovines as well as immunocompromised human patients (Almeida et al., 2006a; Almeida et al., 2006b; Mendonca et al., 2007). Recent data suggest that subgenotyping tools may generate more information about the zoonotic potential of *C. parvum* isolates, although there is still lack of evidences on the usefulness of the generated data for risk assessment (Fayer). *C. hominis*, considered an anthroponotic (human restricted) species, was detected in 1 sample. This species was also reported in water samples and in human stool samples from Portugal (Almeida et al., 2006a; Almeida et al., 2006b; Alves et al., 2003; Lobo et al., 2009). *C. andersoni*, a strictly bovine pathogen, was detected in 16 samples; *C. muris* was detected in only 1 sample suggesting water contamination by rodents.

Curiously, as previously mentioned, no PCR amplification was obtained in drinking water samples. Sensitivity tests were set in the laboratory for the PCR reactions, and the number of 50 cysts or oocysts was our limit of sensitivity. For this reason, the great majority of drinking water samples, with lower parasite load, were out of selection for PCR amplification. Furthermore in the remaining samples subject to PCR, subjected to DNA extraction and PCR, the amplification did not occur. This was also true for a great amount of raw water samples subjected to molecular analysis: 59 samples. This problem has been described by other authors (Jiang et al., 2005), and the most reasonable reason maybe the presence of inhibitor of the polymerase, which was not tested in the present approach or that sufficient amounts of DNA may not be present (empty oocysts or cysts after excystation). Our sensitivity analysis indicates simultaneously a lack of reproducibility in the PCR analysis and the difficulty to achieve amplification in samples with low levels of contamination. The IMS procedure applied over the samples for parasite isolation does not guarantee a complete purity of the sample. Thus, optimization of DNA extraction and amplification protocols is warranted.

The results of this study seem to indicate that the risk assessment for cryptosporidiosis and giardiasis for humans is low in the north of Portugal. However, for this, the condition that whole population has access to the network system for drinking water needs to be fulfilled, and, in this case, this condition is not observed for a very significant segment of the northern population. First of all, according to the national statistics, only 84 % of the population has access to the drinking water system; second, among the population covered by the drinking water distribution systems (84%), a significant segment are provided by water obtained in wells or other origin. Old habits, water prices, and lack of information about the quality of the treated water are the main reasons referred. Taking into account these aspects and the real number of people consuming water for the drinking water systems supply, it is expected that one third of the northern Portuguese population is exposed to the infection by Cryptosporidium and Giardia due to consumption of water. Thus, systematic monitoring of drinking water, livestock, and human biological samples are needed for risk assessment of both diseases. National Health Authorities should consider the urgent implementation of a national monitoring program for microbiological quality of drinking water that includes Cryptosporidium spp. and G. duodenalis analyses. These activities are fundamental steps to understand better the epidemiology of the infection and to allow the implementation of risk analysis models for those infections.

5. Acknowledgments

This work was financially supported by funds of Project 61018, Action Environment and Health, from Calouste Gulbenkian Foundation, and funds of Project PTDC/CVT/103081/2008 from Portuguese Foundation for Science and Technology (FCT).

6. References

Adam, R.D., 2001, Biology of Giardia lamblia. Clin Microbiol Rev 14, 447-475.
Almeida, A.A., Delgado, M.L., Soares, S.C., Castro, A.O., Moreira, M.J., Mendonca, C.M., Canada, N.B., Da Costa, J.M., 2006a, Genotype analysis of Giardia isolated from asymptomatic children in northern Portugal. J Eukaryot Microbiol 53 Suppl 1, S177-178.

Almeida, A.A., Delgado, M.L., Soares, S.C., Castro, A.O., Moreira, M.J., Mendonca, C.M., Canada, N.B., Da Costa, J.M., Coelho, H.G., 2006b, Genetic characterization of Cryptosporidium isolates from humans in northern Portugal. J Eukaryot Microbiol 53 Suppl 1, S26-27.

Alves, M., Xiao, L., Sulaiman, I., Lal, A.A., Matos, O., Antunes, F., 2003, Subgenotype analysis of Cryptosporidium isolates from humans, cattle, and zoo ruminants in Portugal. J Clin Microbiol 41, 2744-2747.

Caccio, S.M., De Giacomo, M., Aulicino, F.A., Pozio, E., 2003, Giardia cysts in wastewater treatment plants in Italy. Appl Environ Microbiol 69, 3393-3398.

Caccio, S.M., De Giacomo, M., Pozio, E., 2002, Sequence analysis of the beta-giardin gene and development of a polymerase chain reaction-restriction fragment length polymorphism assay to genotype Giardia duodenalis cysts from human faecal samples. Int J Parasitol 32, 1023-1030.

Caccio, S.M., Ryan, U., 2008, Molecular epidemiology of giardiasis. Mol Biochem Parasitol 160, 75-80.

Caccio, S.M., Thompson, R.C., McLauchlin, J., Smith, H.V., 2005, Unravelling Cryptosporidium and Giardia epidemiology. Trends Parasitol 21, 430-437.

Castro-Hermida, J.A., Almeida, A., Gonzalez-Warleta, M., Correia da Costa, J.M., Rumbo-Lorenzo, C., Mezo, M., 2007, Occurrence of Cryptosporidium parvum and Giardia duodenalis in healthy adult domestic ruminants. Parasitol Res 101, 1443-1448.

Castro-Hermida, J.A., Garcia-Presedo, I., Almeida, A., Gonzalez-Warleta, M., Correia Da Costa, J.M., Mezo, M., 2008, Contribution of treated wastewater to the contamination of recreational river areas with Cryptosporidium spp. and Giardia duodenalis. Water Res 42, 3528-3538.

Deng, M., Rutherford, M.S., Abrahamsen, M.S., 2004, Host intestinal epithelial response to Cryptosporidium parvum. Adv Drug Deliv Rev 56, 869-884.

Eckmann, L., Gillin, F.D., 2001, Microbes and microbial toxins: paradigms for microbial-mucosal interactions I. Pathophysiological aspects of enteric infections with the lumen-dwelling protozoan pathogen Giardia lamblia. Am J Physiol Gastrointest Liver Physiol 280, G1-6.

Fayer, R., Taxonomy and species delimitation in Cryptosporidium. Exp Parasitol 124, 90-97.

Fayer, R., 2004, Cryptosporidium: a water-borne zoonotic parasite. Vet Parasitol 126, 37-56.

Fayer, R., Morgan, U., Upton, S.J., 2000, Epidemiology of Cryptosporidium: transmission, detection and identification. Int J Parasitol 30, 1305-1322.

Hunter, P.R., Hughes, S., Woodhouse, S., Syed, Q., Verlander, N.Q., Chalmers, R.M., Morgan, K., Nichols, G., Beeching, N., Osborn, K., 2004, Sporadic cryptosporidiosis case-control study with genotyping. Emerg Infect Dis 10, 1241-1249.

Hunter, P.R., Thompson, R.C., 2005, The zoonotic transmission of Giardia and Cryptosporidium. Int J Parasitol 35, 1181-1190.

Jex, A.R., Smith, H.V., Monis, P.T., Campbell, B.E., Gasser, R.B., 2008, Cryptosporidium--biotechnological advances in the detection, diagnosis and analysis of genetic variation. Biotechnol Adv 26, 304-317.

Jiang, J., Alderisio, K.A., Singh, A., Xiao, L., 2005, Development of procedures for direct extraction of Cryptosporidium DNA from water concentrates and for relief of PCR inhibitors. Appl Environ Microbiol 71, 1135-1141.

Lobo, M.L., Xiao, L., Antunes, F., Matos, O., 2009, Occurrence of Cryptosporidium and Giardia genotypes and subtypes in raw and treated water in Portugal. Lett Appl Microbiol 48, 732-737.

McCuin, R.M., Clancy, J.L., 2003, Modifications to United States Environmental Protection Agency methods 1622 and 1623 for detection of Cryptosporidium oocysts and Giardia cysts in water. Appl Environ Microbiol 69, 267-274.

Meinhardt, P.L., Casemore, D.P., Miller, K.B., 1996, Epidemiologic aspects of human cryptosporidiosis and the role of waterborne transmission. Epidemiol Rev 18, 118-136.

Mendonca, C., Almeida, A., Castro, A., de Lurdes Delgado, M., Soares, S., da Costa, J.M., Canada, N., 2007, Molecular characterization of Cryptosporidium and Giardia isolates from cattle from Portugal. Vet Parasitol 147, 47-50.

Monis, P.T., Thompson, R.C., 2003, Cryptosporidium and Giardia-zoonoses: fact or fiction? Infect Genet Evol 3, 233-244.

Muller, N., von Allmen, N., 2005, Recent insights into the mucosal reactions associated with Giardia lamblia infections. Int J Parasitol 35, 1339-1347.

Okhuysen, P.C., Chappell, C.L., Crabb, J.H., Sterling, C.R., DuPont, H.L., 1999, Virulence of three distinct Cryptosporidium parvum isolates for healthy adults. J Infect Dis 180, 1275-1281.

Read, C.M., Monis, P.T., Thompson, R.C., 2004, Discrimination of all genotypes of Giardia duodenalis at the glutamate dehydrogenase locus using PCR-RFLP. Infect Genet Evol 4, 125-130.

Robertson, B., Sinclair, M.I., Forbes, A.B., Veitch, M., Kirk, M., Cunliffe, D., Willis, J., Fairley, C.K., 2002, Case-control studies of sporadic cryptosporidiosis in Melbourne and Adelaide, Australia. Epidemiol Infect 128, 419-431.

Roy, S.L., DeLong, S.M., Stenzel, S.A., Shiferaw, B., Roberts, J.M., Khalakdina, A., Marcus, R., Segler, S.D., Shah, D.D., Thomas, S., Vugia, D.J., Zansky, S.M., Dietz, V., Beach, M.J., 2004, Risk factors for sporadic cryptosporidiosis among immunocompetent persons in the United States from 1999 to 2001. J Clin Microbiol 42, 2944-2951.

Smith, H.V., Caccio, S.M., Tait, A., McLauchlin, J., Thompson, R.C., 2006, Tools for investigating the environmental transmission of Cryptosporidium and Giardia infections in humans. Trends Parasitol 22, 160-167.

Sousa, M.C., Morais, J.B., Machado, J.E., Poiares-da-Silva, J., 2006, Genotyping of Giardia lamblia human isolates from Portugal by PCR-RFLP and sequencing. J Eukaryot Microbiol 53 Suppl 1, S174-176.

Stuart, J.M., Orr, H.J., Warburton, F.G., Jeyakanth, S., Pugh, C., Morris, I., Sarangi, J., Nichols, G., 2003, Risk factors for sporadic giardiasis: a case-control study in southwestern England. Emerg Infect Dis 9, 229-233.

Xiao, L., Fayer, R., 2008, Molecular characterisation of species and genotypes of Cryptosporidium and Giardia and assessment of zoonotic transmission. Int J Parasitol 38, 1239-1255.

Xiao, L., Morgan, U.M., Limor, J., Escalante, A., Arrowood, M., Shulaw, W., Thompson, R.C., Fayer, R., Lal, A.A., 1999, Genetic diversity within Cryptosporidium parvum and related Cryptosporidium species. Appl Environ Microbiol 65, 3386-3391.

Part 5

Management of Environmental Contamination

10

Recycling of Mine Wastes as Ceramic Raw Materials: An Alternative to Avoid Environmental Contamination

Romualdo Rodrigues Menezes[1], Lisiane Navarro L. Santana[2],
Gelmires Araújo Neves[2] and Heber Carlos Ferreira[2]
[1]*Department of Materials Engineering, Federal University of Paraíba*
[2]*Academic Unit of Materials Engineering, Federal University of Campina Grande*
Brazil

1. Introduction

Solid waste management has moved to the forefront of the environmental agenda. Nations are considering restrictions on packaging and controls on products in order to reduce solid waste generation rates. Local and regional governments are requiring wastes to be separated for recycling, and some have even established mandatory recycling targets. However, industrial and everyday activities continue discarding vast amounts of material, some of which contains toxic and environmentally harmful substances. Such substances are not always disposed of in a manner with the avoidance of environmental contamination. Despite the existence of environmental standards, and in spite of the ethical implications of such actions, negligence, cost-cutting and accidents cause contamination of the soil, sediments, water and air. In the last few years, numerous industrial sectors have been mentioned as sources of environmental contamination and pollution due to the enormous quantity of wastes they generate. Mineral extraction and processing are good examples of waste production.

On the other hand, reuse and recycling of waste materials after their potentialities have been detected is considered an activity that can contribute to diversify products, reduce production costs, provide alternative raw materials for a variety of industrial sectors, conserve non-renewable resources, save energy, and especially, improve public health.

The insertion of waste materials into an alternative productive cycle might represent an alternative recovery option, which is interesting from both an environmental and an economical perspective. Recovery and recycling is the best environmental solution to save raw materials and to reduce the amount of industrial waste materials produced, and consequently the contamination of environment.

Considerable research work of our research group has recently focused on the recovery and safe, useful application of waste materials originating from the mining and mineral processing industry. Most of these studies report that such waste can be considered important alternative raw materials to the ceramic industry.

Traditional ceramics, such as bricks, roof and floor tiles, other constructions materials, and technical ceramics, such as porcelain and mullite bodies, are usually highly heterogeneous due to the wide compositional range of the natural clays used as raw materials. Therefore, there is a great incentive to use large amounts of suitable waste products as raw materials. Today it is a well-known fact that some waste materials are similar in composition to the natural raw materials used in the ceramic industry and often contain materials that are not only compatible but also beneficial in the fabrication of ceramics. In view of the huge amounts of non-renewable mineral resources that the ceramic industry consumes, this similarity is of even greater significance.

Studies conducted by the authors observed that kaolin processing wastes and granite sawing waste present a high potential to use as raw material for building materials. Kaolin is an important raw material in various industries, such as the ceramic, rubber, plastic, ink, chemicals, cement, and paper industries. However, the kaolin mining and processing industry generates large amounts of waste. The kaolin industry, which processes primary kaolin, produces two types of waste materials. The first type derives from the first processing step (separation of sand from ore), which represents about 70% of the total waste produced, and is also known as china clay sand. The second type of waste is resulted from the second processing step, which consists of wet sieving to separate the finer fraction and purify the kaolin. Granite processing industry produces large amounts of waste materials worldwide. Granite-sawing waste contains feldspar, quartz and mica as major constituents and metallic dust and lime (used as abrasive and lubricant, respectively) as residual materials, and various studies have studied its recycling in ceramic industry. In this scenario, this chapter will address the use of these mine wastes as raw material for the production of ceramic materials as building materials, ceramic brick and tile, roof tiles, mortars, etc.

2. Recycling and mining wastes

The pollution of ground and surface waters began as soon as industry began producing manufactured goods and wasting liquids and solid matter simultaneously. In the 1930s, industries began to be aware of the eventual danger of their wastes when sent untreated into waterways. It was natural for industry at that time to follow the lead of municipalities in using similar treatments to attempt to resolve their pollution problems. In the World War II there is an accelerated industrial production activity (Nemerow, 2006), and two developments in the post-World War II era led to significant escalation in the problems of managing waste. First, a new phenomenon called "consumerism" emerged. A long period of prosperity, combined with improvements in manufacturing methods led to rapid growth in the number and variety of consumer goods. In addition, new marketing and production practices were introduced, such as planned obsolescence and "throw-away" products. The growth of advertising, along with the electronic media, played an important role in the evolution to our society's current level of overconsumption. The end result was a dramatic increase in the amount and variety of consumer goods—and, hence, wastes (The waste crisis). The second development was the birth of the "chemical age," which resulted in a dramatic change in the composition of the waste stream. The petrochemical industry has grown explosively since that time, yielding a vast array of new synthetic organic compounds, a kind of pollution that had never existed before entered the environment, exhibiting toxicity as well as non-biodegradability (Tammemagi, 1999; Nemerow, 2006).

Radioactivity, petrochemical, and synthetic organic chemicals were largely developed and surfaced in the environment in the 1940s and 1950s. During this period, major environmental problems surfaced with rapid and serious consequences. Hence was born the advent of what was to become the pollution problems of the twentieth century (Nemerow, 2006).

Historically, waste was simply dumped in depressions, ravines, and other handy locales that were close to the population centers producing the waste. Even though recycling was commonly practiced by all households during pre-industrial ages, large-scale recycling programs did not arise until the twentieth century. The first organized programs were created in the 1930s and 1940s, when a worldwide depression limited people's ability to purchase new goods and the outbreak of World War II dramatically increased demands for certain materials. Throughout the war, goods such as nylon, rubber, and various metals were recycled and reused to produce weapons and other materials needed to support the war effort. However, after the War there was a drastically decrease in the recycling efforts (Miller, 2010).

It was not until the environmental movement of the 1960s and 1970s that recycling once again emerged as a popular idea. This movement began in 1962 with the publication of Rachel Carson's book *Silent Spring*, detailing the toxic effects of the chemical DDT on birds and their habitats. The book raised the consciousness of many people about the dangers to the environment from chemicals and other toxins produced by modern industries (Miller, 2010). Thereafter, the increase in the environmental awareness and consciousness required industry to meet tighter environmental standards on a global basis. In many countries, such requirements generally cannot be met by using conventional disposal of residual solid wastes in landfills (Wang et al., 2010). Accordingly, much more emphasis has to be placed on waste reduction and recycling technologies as a necessary first step to reduce to a minimum the extent of the waste treatments to be provided.

In recent years there has been growing concern about the negative impacts that industry and its products are having on both society and the environment in which we live. The concept of sustainability and the need to behave in a more sustainable manner has therefore received increasing attention. With the world's population growing rapidly the consumption of materials, energy and other resources has been accelerating in a way that cannot be sustained (Hester, R. E. & Harrison, 2009).

In this scenario, solid waste management has moved to the forefront of the environmental agenda, with the amount of related activities and concern by citizens and governments worldwide reaching unprecedented levels. Nations are considering restrictions on packaging and controls on products in order to reduce solid waste generation rates. Local and regional governments are requiring wastes to be separated for recycling, and some have even established mandatory recycling targets. Concerns about emissions from incinerators and waste-to-energy plants have resulted in imposition of state-of-the-art air pollution controls. Landfills are being equipped with liners, impervious caps and leachate collection systems, and gas and groundwater is being routinely monitored. As a result, the costs of solid waste management are increasing rapidly (Goumans et al., 1994).

In this context, arise the industrial ecology. Industrial ecology is now a branch of systems science for sustainability, or a framework for designing and operating industrial systems as sustainable and interdependent with natural systems. It seeks to balance industrial production and economic performance with an emerging understanding of local and global ecological constraints (handbook of industrial and hazardous). The idea of industrial ecology is that waste materials, rather than being automatically sent for disposal, should be regarded as raw materials—useful sources of materials and energy for other processes and products (Wang et al., 2006).

Waste management strategies that focus on source reduction and resource recovery, reuse and recycling have proven to be more cost effective over the long run, and they are less damaging to the environment simply because they prevent or minimize waste generation at the source. Disposal and treatment technologies require major long-term investments in capital equipment and have ongoing costs. But in addition, the waste and pollution that are treated and disposed of still persist, posing continuous and future threats to the public and environment (Cheremisinoff, 2003).

Recycling of waste materials will conserve decreasing resources and avoid the environmental and ecological damages caused by their disposal in the environment. Recycling saves energy, preserves natural resources, reduces greenhouse-gas emissions, and keeps toxins from leaking out of landfills.

Successful research and development on using wastes as raw material, is a very complex task. This task comprehends a multidisciplinary approach involving knowledge from different areas, such as materials science, marketing development, performance evaluation and environmental sciences. As a rule, the best application for the waste is the one that will use its true characteristics and properties to enhance the performance of the new product and minimize environmental and health risks. Waste applications should not be made on a preconceived basis. This requires creativity and a wide range of both scientific and technical knowledge and for the best results will require the collaborative work of a multidisciplinary team (Woolley et al., 2000).

However, attention should be given to environmental contamination risk evaluation due to leaching of hazardous components is mandatory. New product must satisfy toxicity leach test criteria. But it is not sufficient. Other environmental impacts like greenhouse gases emission, human toxicity, acidification, energy use, etc. are also important and good technology for recycling frequently allows significant reduction on these impacts (Woolley et al., 2000; Rao, 2006).

On the other hand, public are became more accepting of purchasing manufactured goods with recycled content. Manufacturers recognized this acceptance, that using recycling content in their products developed more innovative ways to use waste material. Manufacturers learned that recycled content yielded economic and marketing benefits, and consumers realized they could buy recycled-content products with confidence (Winkler, 2010).

Mining, alongside agriculture, represents one of man's earliest activities, the two being fundamental to the development and continuation of civilization. In fact, the oldest known mine in the archaeological record is the Lion Cave in Swaziland, which has a radio carbon age of 43 000 years. There Paleolithic humans mined hematite, which they presumably ground to produce the red pigment ochre. Moreover, the dependence of primitive societies on mined products is illustrated by the terms Stone Age, Bronze Age and Iron Age, a sequence of ages that indicate the increasing complexity of the relationship between mining and society (mining and its impact). With time, the use of minerals has increased in both volume and variety in order to meet a greater range of purposes and demand by society, and the means of locating, working and processing minerals has increased in complexity. Today, society is even more dependent on the minerals industry than in the past (Bell & Donnelly, 2006).

Mining is first and foremost a source of mineral commodities that all countries find essential for maintaining and improving their standards of living. Mined materials are needed to construct roads and hospitals, to build automobiles and houses, to make computers and satellites, to generate electricity, and to provide the many other goods and services that consumers enjoy. In addition, mining is economically important to producing regions and countries. It provides employment, dividends, and taxes that pay for hospitals, schools, and public facilities (Committee on Technologies for the Mining Industries et al., 2002).

The consequence of the importance of the mining industry to the world economy is not only the large volume of materials processed but also the large volume of wastes produced. Mine wastes represent the greatest proportion of waste produced by industrial activity. In fact, the quantity of solid mine waste and the quantity of Earth's materials moved by fundamental global geological processes are of the same order of magnitude – approximately several thousand million tonnes per year (Lottermoser, 2007).

Mining, and associated mineral processing and beneficiation, does impact on the environment. Unfortunately, this frequently has led to serious consequences. The degree of impact can vary from more or less imperceptible to highly intrusive and depends on the mineral worked, the method of working, and the location and size of the mine. The environmental impact of mining industry is strongly felt in two areas. The first is the volume of industrial waste, effluents, tailings and sludge. The second serious environmental concern is the emission of carbon dioxide, a major green house gas, which has been implicated in gradual climate change round the world.

Operations of the mining industry include mining, mineral processing, and metallurgical extraction. Mining is the first operation in the commercial exploitation of a mineral or energy resource. Mineral processing or beneficiation aims to physically separate and concentrate the ore mineral, whereas metallurgical extraction aims to destroy the crystallographic bonds in the ore mineral in order to recover the sought after element or compound. All three principal activities of the mining industry produce wastes. Mine wastes are defined as solid, liquid or gaseous by-products of mining, mineral processing, and metallurgical extraction (Lottermoser, 2007).

Mining wastes include overburden and waste rocks excavated and mined from surface and underground operations. Mining wastes are heterogeneous geological materials and may consist of sedimentary, metamorphic or igneous rocks, soils, and loose sediments. As a consequence, the particle sizes range from clay size particles to boulder size fragments. The physical and chemical characteristics of mining wastes vary according to their mineralogy and geochemistry, type of mining equipment, particle size of the mined material, and moisture content (Lottermoser, 2007).

Mineral processing encompasses unit processes for sizing, separating and processing minerals, including comminution, sizing, separation, dewatering, some types of chemical processing. Processing wastes the portions of the crushed, milled, ground, washed or treated resource deemed too poor to be treated further. The physical and chemical characteristics of processing wastes vary according to the mineralogy and geochemistry of the treated resource, type of processing technology, particle size of the crushed material, and the type of process chemicals.

Metallurgical wastes are the residues of the leached or smelted resource deemed too poor to be treated further, and are generated by hydrometallurgical extraction and electro- and pyrometallurgical processes.

Mine wastes result from the extraction of metalliferous and non-metalliferous deposits. In the case of metalliferous mining, high volumes of waste are produced because of the low or very low concentrations of metal in the ore. In fact, mine wastes represent the highest proportion of waste produced by industrial activity, billions of tonnes being produced annually. Such wastes can be inert or contain hazardous constituents but generally is of low toxicity. The chemical characteristics of mine waste and waters arising from them depend upon the type of mineral being mined, as well as the chemicals that are used in the extraction or beneficiation processes. Because of its high volume, mine wastes historically have been disposed of at the lowest cost, often without regard for safety and often with considerable environmental impacts (Bell & Donnelly, 2006).

Usually the metalliferous and non-metalliferous wastes (from mining and processing operations) are placed in the postmining topography. In the case of mountain top removal and contour mining methods, waste materials are often used to fill adjacent canyons or hollow areas. When associated with canyon fills, these anthropogenic land forms may be flat or gently sloping on top, but often have steep side slopes and tend to be very erosive. Also, because of the nature of the material (i.e., unconsolidated, non-homogeneous) water penetration can cause instability thus enhancing mass wasting and the formation of seeps containing high levels of various elements that could impact down slope sites (Marcus, 2007).

Another very important concern to the mining industry is particulate matter, which is emitted in relatively large amounts in almost all aspects of mining operations (mining environmental). Particulates can affect human health adversely, as well as damage animals and crops. At high enough levels, particulates can contribute to chronic respiratory illnesses such as emphysema and bronchitis and have been associated with increased mortality rates from some diseases. In addition, particulate matter may cause irritation of the eyes and throat, and it can impair visibility.

As processing technologies move toward finer and finer particle sizes, dust and fine particles produced in the mineral industry are becoming an important consideration Fine particles and dust can represent a health hazard, an environmental concern, and an economic loss. The amount of waste dust and fine particles is increasing significantly as more rock is mined and processed. Research should be focused on minimizing the generation of unwanted fine particles and dust or on using these materials as viable by-products.

Moreover, large volumes of water slurries containing fine particles are produced by all types of mining facilities. The management of these slurries as they are dewatered and disposed of can present significant environmental issues. Whether slurries are produced as tailings from milling operations, spoils in coal mining, or as clay slimes in the phosphate industry, they are often slow and difficult to dewater and dry because of their colloidal nature (Committee on Technologies for the Mining Industries et al., 2002).

There are other problems faced by the mining industry, as closure and reclamation of dump-leaching and heap-leaching operations and tailings impoundments. Upon the cessation of production, dump-leaching and heap-leaching piles and tailings impoundments must be closed in an environmentally sound manner. Depending on the chemical characteristics of the wastes and reagents used, as well as on atmospheric precipitation rates, piles and tailings may contribute poor-quality seepage or runoff to surface and/or groundwater through the release of residual solution or from infiltration of or contact with atmospheric precipitation. The released solution may be acidic or may contain cyanide or other contaminants, such as selenium, sulfates, radionuclides, or total dissolved solids (Committee on Technologies for the Mining Industries et al., 2002).

However, not all mine wastes are problematic wastes and require monitoring or even treatment. Many mine wastes do not contain or release contaminants, and pose no environmental threat. In fact, some waste rocks, soils or sediments can be used as raw materials for a series of industries. , and a few are suitable substrates for vegetation covers and similar rehabilitation measures upon mine closure.

Therefore, the development of innovative, environmentally friendly technologies will be extremely important. Minimizing waste generation and using wastes to produce useful by-products while maintaining economic viability must be a goal for new technologies. For instance, the processing of metallurgical wastes and recovery of valuable components and in some cases converting them into useful compounds not only help to reduce pressure on ponds and landfills but also it, at least in part, offsets the cost of environmental protection (Rao, 2006).

Sustainable development is a concept which attempts to shape the interaction between the environment and society, such that advances in wellbeing are not accompanied by deterioration of the ecological and social systems which will support life into the future. The management of mining and minerals processing wastes is therefore a fundamental sustainable development issue.

The reduction, reuse, recycling and treatment of mining and minerals processing waste are increasingly receiving greater research and development attention for their contribution to improving the sustainability of the minerals industry (Franks et al., 2011) Recent trends outlined the new mining culture: management of resources on a local basis (water, building materials, etc.), prevention of dissipative uses by improved recycling, and promotion of efficiency in production processes (better recovery means less pollution), etc.

It is very important to be aware of the fact that almost the same kind of material, depending on its characteristics, could be regarded as a waste (that may have to be treated) or a secondary raw material (environmental aspects of construction). This implies that close cooperation between society players (in both public and private sectors) dealing with waste recycling technology to replace raw materials by mining wastes.

The quantity of industrial minerals recycled or reused in some way is still minor compared to the total global consumption of industrial minerals obtained by mining and quarrying (resource recovery). However, manufacturers began to look at recycled waste as a more reliable and cost-effective supply source for raw material, and altered existing products or developed new ones to better use recycled products (Winkler, 2010). In this sense mining wastes have being used in resin cast products, glass, ceramics, glazes as well as building and construction materials.

Thus, mining waste can no longer be considered solely a useless material, but, it is must first analyzed (scientifically and technically) the potential usefulness of the wastes as alternative raw materials, reducing the demand for virgin materials and the environmental impacts associated with extraction and processing of virgin resources.

3. Granite sawing waste and kaolin processing waste: Alternative ceramic raw materials

The industry of the ornamental granite is included in the natural stone industry sector, more specifically in the sub-sector of the ornamental rocks comprising the extraction and processing of rocks for ornamental applications. This economical activity represents an important sector in the worldwide economy. However, a less known aspect of the exploration of the ornamental rocks is the great volume of produced residues, specifically, solids (generated during the extraction) and sludges (produced during the transformation process) (Torres et al., 2009).

After mining, the granite blocks are submitted to primary dressing, which consist of sawing (Figure 1 depicts the sawing operations), to obtain semi finished pieces such as plates and strips. This is followed by secondary dressing in which the sectioned pieces undergo polishing and surface finishing. During primary dressing, an estimated loss of 20–35% of the blocks occurs in the form of powder. This leftover powder is removed in a mixture with water and other residual materials such as metallic dust and lime, used as abrasive and lubricant, respectively, producing a sludge.

Hence, granite waste, as described elsewhere (Menezes et al., 2002a, 2002b; Torres et al., 2004; Menezes et al., 2005) is basically composed of quartz, mica, potassium feldspar ($KAlSi_3O_8$), sodium feldspar ($NaAlSi_3O_8$), iron (or iron oxide and iron hydroxides) and lime (or calcite ($CaCO_3$)) and has a large particle size distribution.

This material has been deposited as dry particles with large particle size range or as fine particles in aqueous environment, generally deposited by sedimentation. It is also common to deposit filter-pressed sludge in surface landfill. Although not considered dangerous, incorrectly planned deposition of these residues can cause accidents and environmental impact like, for instance, the increase of the turbidity of the courses of water. The dried mud is easily dragged by the wind and becomes harmful to humans and animals through its inspiration or to plants when deposited on their leaves. The sludge generated during the processing of stone have no specific practical applications and have been managed as waste. Figure 2.

Thus, the search for new recycling technologies is of high technological, economic and environmental interest. In this regard, interesting opportunities are found in the traditional ceramics industry, particularly the sector devoted to the fabrication of building products. Natural raw materials used in the fabrication of clay-based ceramic products show a wide range of compositional variations and the resulting products are very heterogeneous.

Fig. 1. Sawing operation of granite blocks.

Granite sludge is a mixture of debris residue of cut granite rocks with wear remains of cutting steel blades, abrasive metallic shot, and hard materials from the polishing bricks. In the common granite cutting practice, the abrasive metallic shot is dispersed in $Ca(OH)_2$ aqueous slurry for cooling. This slurry is continuously pumped and wets all around the granite block slits (Figure 1).

Therefore, such products can tolerate further compositional fluctuations and raw material changes, allowing different types of wastes to be incorporated into ceramic tiles and bricks. Various studies (Menezes et al., 2002a, 2005, 2007, 2008a) have demonstrated the viability of using granite sawing waste in the production of building materials.

Studies of our research group demonstrated that is possible to incorporate high amounts granite sawing waste in ceramic formulations for the production of ceramic bricks and roof tiles. Figure 3 illustrate the mechanical behavior of ceramic bodies, fired at 1000°C, with the rise of waste content. Small additions of these wastes, up to 20%, improved the mechanical performance of the bodies. Granite wastes present a non-plastic character and, therefore, it was also observed that they can play an important role as plasticity-controllers during fabrication. Figure 4 displays ceramic bodies produced incorporating granite sawing.

For the production of ceramic bricks, the predominant raw material used is mineral clay. Any good brick clay should have low shrinkage and low swelling characteristics, consistent firing color, and a relatively low firing temperature, but at the same time produce an adequately dry and fire-strength brick. The guiding rule of choice on wastes and by-products must rest on their compatibility with the original (host) raw material being used, whereas they must not degrade the final product by focusing simply on making it a repository for wastes (Insam & Knapp, 2011). Granite sawing waste reaches all the requirements to be a versatile raw material for the production of ceramic brick and roof tiles.

Granite sawing waste can also be used in the production of soil-lime bricks. Figure 5 shows a soil-lime brick containing granite waste and a construction developed using this kind of brick.

In ceramic technology there is a large range o firing temperatures. After 1100°C granite sawing waste can be classified as fluxes, as they have the potential to act as glassy phase formers during the sintering process, improving the sinterability of the clay material. The effect of small additions of such rejects in compositions for the production of ceramic tiles has been investigated (Menezes et al., 2005, 2008a), and it was observed that the final properties of the fired products do not change drastically. According to the composition to be manufactured it is possible incorporate high amounts of waste, up to 50%.

Figure 6 depicts the water absorption and mechanical behavior of ceramic tiles (produced with different composition) fast fired at 1175°C, containing distinct amounts of different granite wastes. It can be observed that there is no a directly correlation between these properties and waste content. This is due to the fact that each granite wastes present particular characteristics, as amount of fluxes and particle size distribution. But also, because the influence of granite wastes will be close associated with the characteristics (and amounts) of the other raw materials used in the formulation of ceramic masses. Thus, it is possible incorporated 38% of waste and reached high mechanical performance, and in other situation using 21% of waste the modulus of rupture achieved was just above the strength limit required.

The current optimization procedure for developing ceramic compositions using waste materials consists of an experimental rather than a comprehensive approach. In general, the approach involves selecting and testing a first trial batch, evaluating the results, and then adjusting the mixture's proportions and testing further mixtures until the required

Fig. 2. Disposal of granite sawing waste in the environment: in aqueous environment and in open air

properties are achieved. The conventional method of optimization is time consuming and does not allow the global optimum to be detected, particularly due to interactions among the variables. In contrast, statistical design methods are rigorous techniques both to achieve desired properties and to establish an optimized mixture for a given constraint, while minimizing the number of trials. This methodology when applied in the recycling of wastes in the ceramic technology has led to greater efficiency and confidence in the results obtained and has simultaneously optimized the content waste materials with a minimum of experiments.

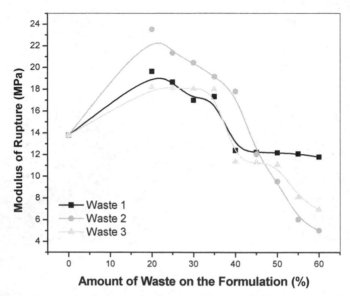

Fig. 3. Modulus of rupture of ceramic bodies fired at 1000°C (formulations for the production of ceramic bricks and roof tiles)

Fig. 4. Ceramic bodies containing granite sawing waste

Fig. 5. Soil-lime brick containing granite sawing waste and a construction developed using this kind of brick.

In the development and manufacture of ceramics using waste materials, the properties of fired bodies are basically determined by the combination of raw materials and process parameters. When the processing conditions are kept constant, a number of properties of dried and fired bodies are basically determined by the combination (or mixture) of raw materials. This is the basic assumption in the statistical design of mixture experiments to obtain a response surface using mathematical and statistical techniques. To this end, it is necessary first to select the appropriate mixtures from which the response surface might be calculated. Then, form the calculated response surface, the property value of any mixture can be predicted based on the changes in the proportions of its components. In this sense authors had applied this mathematical tool in several studies developing ceramic formulations containing high amount of granite wastes with great efficiency and a minimum of experiments.

Figure 7 shows the calculated response surface plots and their projections onto the composition triangle for the modulus of rupture ceramic bodies containing granite sawing waste after firing at 1150ºC. According to the Figure it is possible to quickly realize the real influence of granite waste on the property of ceramic body and also optimize raw materials composition, increasing the content of waste and, in the same time, improving the performance of the final body. In this case in particular it is clear that the maximum value of the modulus of rupture was achieved when using at around 50% of granite sawing waste.

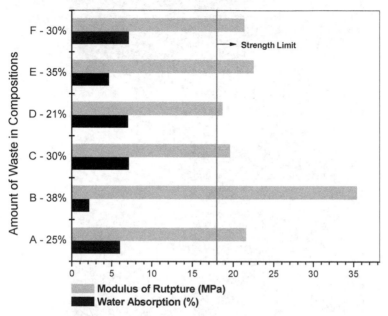

Fig. 6. Water absorption and modulus of rupture of ceramic tiles fired at 1175°C containing granite sawing waste

Kaolin is an important raw material in various industries, such as the ceramic, rubber, plastic, ink, chemicals, cement, and paper industries. However, the kaolin mining and processing industry generates large amounts of. The kaolin industry, which processes primary kaolin, produces two types of wastes. The first type derives from the first processing step (separation of sand from ore, generally by wet sieving). The other type of waste results from the second processing step, which consists of a wet sieving to purify the kaolin. Figure 8 display the second step of the kaolin processing, the sedimentation tank.

Traditionally, these wastes have been disposed of in landfills and often dumped directly into ecosystems without adequate treatment. This can seriously damage the environment through soil and water contamination, and is potentially harmful to flora, fauna and human health. Figure 9 depicts the dispose of these wastes directly in open air. Nowadays, because of more stringent environmental laws and the market's increasing demand for environmentally friendly products, manufacturers are concerned with developing studies aimed at reducing the environmental impact of these wastes. Thus, possible reuse or recycling alternatives should be investigated and implemented.

Physical and chemical characterizations of kaolin processing wastes can be found elsewhere (Menezes et al., 2007, 2008b). According to those reports, kaolin wastes are basically composed of kaolinite ($Al_2Si_2O_5(OH)_4$), mica ($KAl_2(Si_3Al)O_{10}(OH)_2$) and quartz ($SiO_2$) and has a very large particle size distribution. The composition and particle size distribution of the two types of waste are very different. The first one contain a high amount of quartz and coarse particle size (reaching particles of 5, 10mm), while the second one, presents a high amount of kaolinite and a large particle size distribution with a high mount of fine particles (because of this the second waste are known as fine waste and the first waste as coarse waste).

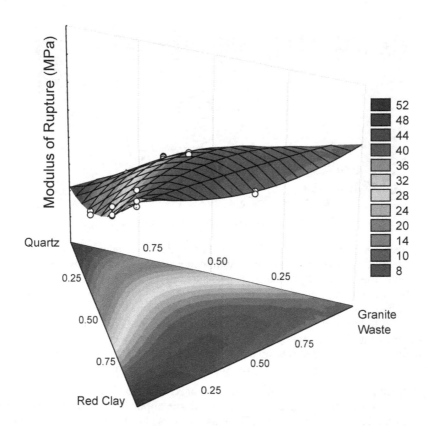

Fig. 7. Response surface plots and their projections onto the composition triangle for modulus of rupture of ceramic bodies containing granite sawing waste after firing at 1150°C

Mineral fillers are used in a wide range of commodities such as paper, paint, plastics, membranes, ceramics, plasterboard, geo-textiles, rubber, pet food, chicken feed, electrical cables and several construction materials. Such fillers are marketed at a relatively high cost, as the total production costs. However many applications such as ceramics and some construction materials, do not require fillers of such a high grade. In this sense, the fine kaolin processing waste was studied by our research group to be used as filled in mortars for the construction industry. The waste replaced the lime and the cement in several mortars formulations presenting very interesting results. Figure 10 shows the compression strength of mortars, which had the lime replaced by kaolin processing waste (fine waste). Addition of waste improved the performance of the mortar due to the filler effect of the material.

Fig. 8. Sedimentation tank used in the second step of kaolin processing

Kaolin processing waste after be fired display pozzolanic activity (capacity to react chemically with the lime and form hydrated calcium silicates, phases similar to those produced with the cement hydration). Thus, calcinations of this waste can improve its applicability and incorporation in construction materials. The compression strength of mortars that had part of cement replaced by fired kaolin waste (50% of coarse and 50% of fine) is depicted in Figure 11. This result illustrates the potential of this waste as agglomerate material after firing. The increase in the compression strength when using waste in natural is due to the filler affect. Because of the high fineness, their particles can fill the voids between the cement particles, increasing the soil density and strength.

Kaolin processing waste (fine waste) can also be used in other ceramic industries. Studies (Menezes et al., 2008b, 2009a, 2009b) have pointed up its application in production of porous ceramics, mullite bodies and porcelains. The waste acts as alternative raw material replacing part of the kaolin and of the quartz used in the formulation. Bodies obtained presented high strength and excellent performance, similar to those of bodies produced using conventional raw materials. The outstanding performance of the produced bodies was close associated with the development of high amount of mullite, as a consequence of the presence of fluxes and kaolinite on the waste (Figure 12).

The potential use of granite waste in combination with kaolin processing waste to produce ceramic bodies was also investigated by our research group. Regression models used to optimize the waste content in ceramic compositions displayed that ceramic bricks containing up to 40% of wastes (kaolin waste + granite waste) can be manufactured without trouble and the final bricks presented physical and mechanical properties similar to those of conventional bricks. Figure 13 illustrates this potential using the surface plot projection onto the composition triangle. The area highlighted on the triangle indicates the possible compositions (according to limitations imposed) to be used for ceramic brick production.

Fig. 9. Piles of wastes directly disposed in the environment

Use of kaolin waste in association with granite waste was also very efficient for the production of ceramic tiles containing high amount of wastes, at around 60%. Figure 14 displays the synergism of the wastes, while the granite waste improves the mechanical behavior of the body, the kaolin waste act in a manner similar to the clay. The combination of both wastes make possible improves the performance of the body and save clay material, using high amount of kaolin waste and granite waste.

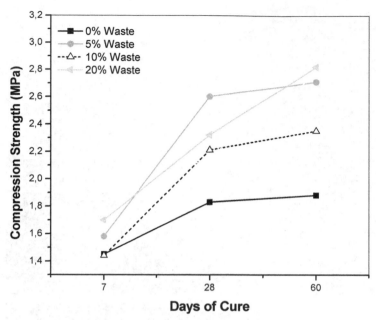

Fig. 10. Compression strength of mortar (cement:lime:sand), which had the lime replaced by kaolin processing waste

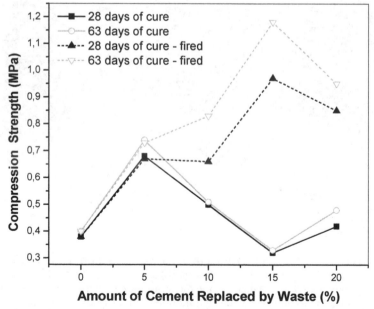

Fig. 11. Compression strength of mortar (cement:sand), which had part of the cement replaced by kaolin processing waste

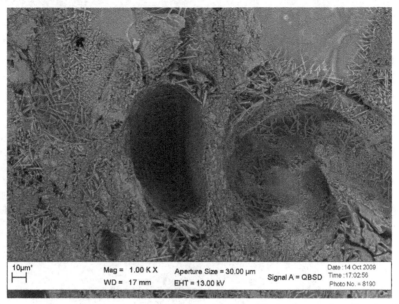

Fig. 12. Scanning electron microscopy micrograph of mullite body produced using kaolin processing waste (fine waste)

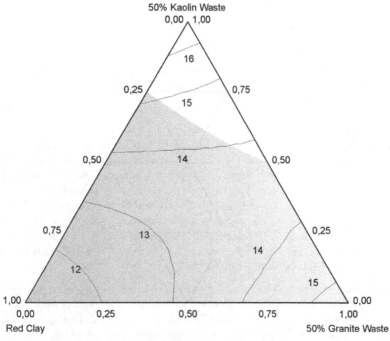

Fig. 13. Surface plot projection onto the composition triangle of compositions containing kaolin processing and granite sawing wastes

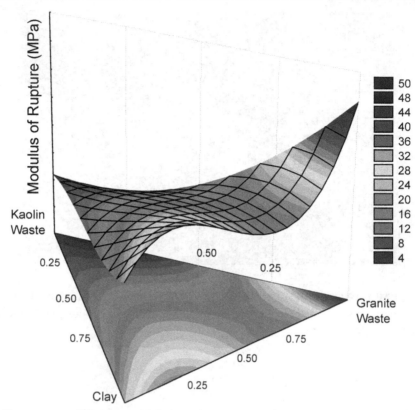

Fig. 14. Response surface plots and their projections onto the composition triangle for modulus of rupture of ceramic bodies containing kaolin processing and granite sawing wastes after firing at 1100°C

While recycling of low added-value residual materials constitutes a present day challenge in many engineering branches, attention has been given to low-cost building materials with similar constructive features as those presented by materials traditionally employed in civil engineering. Developing countries usually face grave housing deficits. Aiming at lowering costs, scientific attention has been given to non-conventional building materials with similar features as those presented by construction materials traditionally used in civil engineering. Quest for such surrogate materials can be two-fold interesting as (i) it may help to reduce dwelling deficits (particularly in developing countries) inasmuch as cheaper houses become economically feasible and (ii) it can be environmentally friendly as low-value wastes can be recycled or exploited. (Ashworth & Azevedo, 2009).

On the other hand, these results on recycling of granite sawing and kaolin processing wastes highlight that these wastes recycling contributes to diversify products, reduces production costs, provides alternative raw materials for a variety of industrial sectors, conserves non-renewable resources, saves energy and, improves public health. Thus, upgrading these wastes to alternative ceramic raw materials has become interesting, not only technically, but also environmentally and socially.

4. Conclusion

Studies of our research group displayed that kaolin processing and granite sawing wastes can serve as alternative raw materials for the production of ceramic materials, and that not only construction materials (as ceramic bricks and tiles, roof tiles, mortars, etc.) but also ceramics like porcelain, mullite bodies, membranes, etc. The correct use of the wastes produces lower firing temperature or improves the performance of the final bodies. Correct characterization, physically and microstructurally, and application of mathematical tools permit incorporation of high amounts of waste in ceramic formulations, exceeding 50% of the raw material used in the composition. Our results highlighted that the use of mine wastes in the production of building materials can be successfully carried out, allowing the reduction of both the consumption of natural resources and the cost of waste disposal while protecting the environment.

5. Acknowledgment

The authors gratefully acknowledge the Brazilian Council of Research, CNPq (Proc. 577201/2008-5 and 303388/2009-9) for financial support.

6. References

Ashworth, G. S. & Azevedo, P. (Ed(s)) (2009). *Agricultural Waste*, Nova Science Publishers Inc., ISBN 978-1-61728-132-7, New York.

Bell, F. G. & Donnelly, L. J. (2006). *Mining and its Impact on the Environment*, Taylor & Francis, ISBN10 0–203–96951–0, New York.

Cheremisinoff, N. P. (2003). *Handbook of Solid Waste Management and Waste Minimization Technologies*, Butterworth-Heinemann, Elsevier Science, ISBN 0-7506-7507-1, Burlington.

Committee on Technologies for the Mining Industries, National Materials Advisory Board, Board on Earth Sciences and Resources, Committee on Earth Resources, National Research Council (2002). *Evolutionary and Revolutionary Technologies for Mining*. National Academy Press, ISBN 0-309-07340-5, Washington, D.C.

Franks, D. M.; Boger D. V.; Côte, C. M. & Mulligan, D. R. (2011). Sustainable development principles for the disposal of mining and mineral processing wastes. *Resources Policy* Vol. , pp. 114–122, ISSN 03014207

Goumans, J. J. J. M.; van der Sloot, H. A. & Aalbers, Th. G. (1994). *Environmental Aspects of Construction with Waste Materials*, Elsevier, ISBN 0-444-81 853-7, Amsterdam.

Hester, R. E. & Harrison, R. R. (2009). *Electronic Waste Management*, Royal Society of Chemistry (RSC), ISBN 978-0-85404-112-1, Cambridge.

Insam, H & Knapp, B. A. (Ed(s)) (2011). *Recycling of Biomass Ashes*, Springer-Verlag Berlin Heidelberg, ISBN 978-3-642-19353-8, New York.

Lottermoser, B. G. (2007). *Mine Wastes Characterization, Treatment, Environmental Impacts*, Second Edition, Springer-Verlag Berlin Heidelberg, ISBN-10 3-540-48629-1, New York.

Marcus, J. J. (Ed) (1997). *Mining Environmental Handbook: Effects of Mining on the Environment and American Environmental Controls on Mining*. Imperial College Press, ISBN 1-86094-029-3, London.

Menezes, R. R.; Almeida, R. R.; Santana, L. N. L.; Neves, G. A.; Lira, H. L. & Ferreira, H. C. (2007). Análise da co-utilização do resíduo do beneficiamento do caulim e serragem de granito para produção de blocos e telhas cerâmicos. *Cerâmica*, Vol. 53, pp. 192-199, ISNN 0366-6913 (www.scielo.br).

Menezes, R. R.; Brasileiro, M. I.; Gonçalves, W. P.; Santana, L. N. L.; Neves, G. A.; Ferreira, H. S. & Ferreira, H. C. (2009b). Statistical Design for Recycling Kaolin Processing Waste in the Manufacturing of Mullite-Based Ceramics. *Materials Research*, Vol. 12, No. 2, pp. 201-209, ISSN 1516-1439.

Menezes, R. R.; Farias, F. F.; Oliveira, M. F.; Santana, L. N. L.; Neves, G. A.; Lira, H. L. & Ferreira, H. C. (2009a). Kaolin processing waste applied in the manufacturing of ceramic tiles and mullite bodies. *Waste Management & Research*, Vol. 27, pp. 78–86, ISSN 0734–242X.

Menezes, R. R.; Ferreira, H. S.; Neves, G. A.; Lira, H. L.; Ferreira, H. C. (2005). *Journal of the European Ceramic Society*, Vol. 25, pp. 1149–1158, ISSN 0955-2219.

Menezes, R. R.; Malzac Neto, H. G.; Santana, L. N. L.; Lira, H. L.; Ferreira, H. S. & Neves, G. A. (2008a). *Journal of the European Ceramic Society*, Vol. 28, pp. 3027–3039, ISSN 0955-2219.

Menezes, R. R.; Neves, G. A. & Ferreira, H. C. (2002b). O estado da arte sobre o uso de resíduos como matérias-primas cerâmicas alternativas. *Revista Brasileira de Engenharia Agrícola e Ambiental*, Vol. 6, No.2, pp. 303-313, ISSN 1415 4366.

Menezes, R. R.; Neves, G. A.; Ferreira, H. C. & Lira, L. L. (2002a). Recycling of granite industry waste from the northeast region of Brazil. *Environmental Management and Health*, Vol. 13, No. 2, pp. 134-141, ISSN 0956-6163.

Menzes, R. R.; Brasileiro, M. I.; Santana, L. N. L.; Neves, G. A.; Lira, H. L & Ferreira, H. C. (2008b). Utilization of kaolin processing waste for the production of porous ceramic bodies. *Waste Management & Research*, Vol. 26, pp. 362–368, ISSN 0734–242X.

Miller, D. A. (2010). *Garbage and Recycling*, Lucent Books, ISBN 978-1-4205-0147-6, Farmington Hills.

Nemerow, N. L. (2006). *Industrial Waste Treatment: Contemporary Practice and Vision for the Future*, Elsevier Science & Technology Books, ISBN 0123724937, Amsterdam.

Rao, S. R. (2006). *Resource Recovery and Recycling from Metallurgical Wastes*, Elsevier, ISBN 0-08-045131-4, Amsterdam.

Tammemagi, H. (1999). *The Waste Crisis: Landfills, Incinerators, and the Search for a Sustainable Future*. Oxford University Press, ISBN 0-19-512898-2, New York.

Torres, P.; Fernandes, H. R.; Agathopoulos, A.; Tulyaganov, D. U. & Ferreira, J. M. F. (2004). Incorporation of granite cutting sludge in industrial porcelain tile formulations. *Journal of the European Ceramic Society*, Vol. 24, pp. 3177–3185, ISSN 0955-2219.

Torres, P.; Fernandes, H. R; Olhero, S. & Ferreira, J. M. F. (2009). Incorporation of wastes from granite rock cutting and polishing industries to produce roof tiles. *Journal of the European Ceramic Society*, Vol. 29, pp. 23–30, ISSN 0955-2219.

Wang, L. K.; Hung, Y. T. & Shammas, N. K. (2010). *Handbook of Advanced Industrial and Hazardous Wastes Treatment*, CRC Press, ISBN 978-1-4200-7219-8, Boca Raton.

Wang, L. K.; Hung, Y. T.; Lo H. H. & Yapijakis, C. (2006). *Handbook of Industrial and Hazardous Wastes Treatment*, Second Edition, Marcel Dekker, ISBN 0-203-02651-9, New York.

Winkler, G. (2010). *Recycling Construction & Demolition Waste - A Leed-Based Toolkit*, McGraw-Hill, ISBN 978-0-07-171339-9, New York.

Woolley, G. R.; Goumans, J. J. J. M. & Wainwright, P. J. (Ed(s)) (2000). *Waste Materials in Construction: The Science and Engineering of Recycling for Environmental Protection*, Pergamon/Elsevier, ISBN 0-08-043790-7, Amsterdam.

11

Risk Assessment and Management of Terrestrial Ecosystems Exposed to Petroleum Contamination

M. S. Kuyukina[1,2], I. B. Ivshina[1,2], S. O. Makarov[2] and J. C. Philp[3]
[1]*Institute of Ecology and Genetics of Microorganisms, Russian Academy of Sciences, Perm*
[2]*Perm State University, Perm*
[3]*Science and Technology Policy Division, Directorate for Science Technology and Industry, OECD*, Paris*
[1,2]*Russia*
[3]*France*

1. Introduction

Risk assessment as part of a strategy for dealing with contaminated land is becoming the norm internationally. The term covers both human health and ecological risk assessment. Risk assessment may be defined as the "characterisation of the potential adverse health effects of human exposures to environmental hazards" (human health risk assessment) or the "process of estimating the potential impact of a chemical or physical agent on a specified ecological system under a specific set of conditions" (ecological risk assessment) (Markus & McBratney, 2001).

The risk assessment movement was probably born as a result of experiences with its forerunner, the multifunctionality approach, which was particularly prevalent in the Netherlands. The Dutch example serves as our introduction to this change of policy from multifunctionality to suitability for use, or functionality. The situation was elegantly described by Honders et al. (2003).

In the Netherlands in the early 1980's, all contaminated sites had to be fully excavated and remediated to the level of the reference values (natural contaminant concentrations in soil), in order that the land could be used for a variety of purposes. This very rigorous stance was based upon the perception that the total national remediation costs would be in the order of Euro 0.5 billion. Remediation funds were largely provided by the national government. By the end of the eighties, it had become clear that total remediation costs were going to be in the order of Euro 50 billion if all sites were to be cleaned up to this very high standard.

The Soil Protection Act in the Netherlands initiated the move towards a risk-based approach to site remediation. The concept of "multifunctionality" was replaced by the concept of

---

* The opinions expressed and arguments employed herein are those of the author(s) and do not necessarily reflect the official views of the OECD or of the governments of its member countries.

"functionality", or "suitability for use", or "fitness for use". Quite clearly if a contaminated site was to be re-developed as, say, a car park that would be covered with concrete or tar, then the exorbitant cost of decontamination to reference values was unwarranted. On the other hand, if the site was to be re-developed for a purpose that would involve the exposure of humans to contaminants (for example, a housing development or kintergarden) then removal of contaminants to a higher level, and therefore higher cost, was justified.

In this way a risk assessment would determine the clean-up standard and could also be used in the selection of a remedial technology for the site (the so-called risk-based remedial design). In the eighties, *ex-situ* treatment technologies were still in their infancy. Incineration of contaminated soil, whilst controversial, ensures destruction of the contaminants and was therefore a reasonable option for sites that required to be cleaned to a very high standard. Unfortunately incineration also destroys the soil itself. There has always been concern about the destruction of soil, and the European Union has acknowledged this with the proposal for a directive for the protection of soil. The following is taken from that proposal (Commission of the European Communities, 2006).

> *"Soil is a resource of common interest to the Community, although mainly private owned, and failure to protect it will undermine sustainability and long term competitiveness in Europe. Moreover, soil degradation has strong impacts on other areas of common interest to the Community, such as water, human health, climate change, nature and biodiversity protection, and food safety".*

A common technique internationally for dealing with contaminated land at that time was landfill, which became known as "dig and dump". However, it is recognised that this is not a treatment option; rather it simply moves the problem somewhere else as the anaerobic environment of landfill is not conducive to the destruction of organic contaminants. Besides, landfill is under strong scrutiny as sites suitable for development for landfill become rarer. Even in a country like Australia, with a large land mass and low population, there are good reasons to consider the available supply of landfill to be a scarce resource that should be used conservatively (Pickin, 2009). A country with quite the opposite conditions is Japan, where there is limited space and high population density. In Japan, it is becoming increasingly difficult to obtain public acceptance to install waste disposal facilities, such as landfill sites, due to a rising pressure on land use and growing public concern over environmental and health protection (Ishizaka & Tanaka, 2003). There has been legislation developed in many countries aimed at maximising the efficiency of use of landfill sites, and dumping contaminated soil in them does not represent a good use of space.

Landfill has typically been the least expensive option, compared with, say, bioremediation and soil washing. However, with the arrival of Landfill Tax, the costs of alternatives to landfill disposal become more comparable. It was predicted in the UK that, as more experience was gained with alternative technologies, costs should fall, helping to encourage new cost-effective remediation approaches (Day et al., 1997). That prediction has come true.

The risk assessment concept favours technologies such as bioremediation. Bioremediation has had difficulty competing with other remedial technologies because of some uncertainties, such as remedial target end points and the time required (Diplock et al., 2009). It has therefore been more difficult to establish engineering parameters (Philp et al., 2005a). With risk assessment, and the inherently less rigid outcomes for remediation that may be derived from the assessment, bioremediation technologies have become more attractive. Bioremediation technologies are now deployed internationally and are cost-competitive.

Scope of the chapter

Figure 1 shows the overall contaminated land management process (modified from DEFRA and Environment Agency, 2004, with all the steps other than risk assessment in grey for clarity). This chapter will be confined to the left column, which is the risk assessment part of the management process, although some comments will be made about risk-based remedial design i.e. remedial options appraisal. As can be seen from the flowchart, risk assessment in the context of contaminated land management is an iterative process, during which more site data may or may not be required depending on the complexity of the contamination problem. Most common problems associated with terrestrial petroleum contaminations are old petrol station sites in urban areas and accidental crude oil-spillage sites along cross-country pipelines, which are complex, with multiple contaminants, and typically require a remediation treatment train. The greater the number of pollutant linkages, the greater the requirement for iteration.

2. Drivers for contaminated site remediation

In the European context the three major drivers for contaminated site clean-up are:

1. Direct regulatory intervention;
2. The need for the development of urban industrial areas ("brownfield sites") (van Hees et al., 2008);
3. National, mainly state-funded, programmes ("orphan sites").

The most widely accepted definition of brownfields is that of the US EPA (1997), where they are described as: "abandoned, idled, or under-used industrial and commercial facilities where expansion or redevelopment is complicated by real or perceived environmental contamination". The widespread problem of brownfield sites is the result of two concurrent factors (Alberini et al., 2005):

1. The 1970's saw the down-sizing of US and Western European manufacturing, with many factory closures;
2. The passage of environmental legislation, especially based on the polluter pays principle, whereby parties were identified with the responsibility for the clean-up of contaminated sites.

It has been said that the most significant driver of the regeneration of contaminated sites in the UK is the development process, especially for brownfeld sites (Luo et al., 2009). This is likely to be the case in relatively small, but densely populated countries with a high demand for housing provision. It has also been the case in the US and Canada (de Sousa, 2003).

Scarcity of land, particularly for house building, has raised the political profile of brownfield site redevelopment, and as a result contaminated land has gradually risen up the political agenda (Catney et al., 2006). The development of brownfield sites helps prevent the use of green sites for housing, and also promotes economic growth in inner cities, and is therefore a potentially important component of sustainable growth.

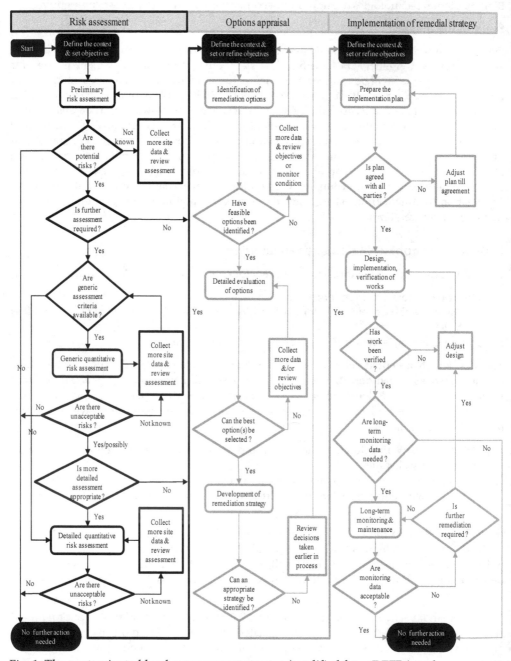

Fig. 1. The contaminated land management process (modified from DEFRA and Environment Agency, 2004).

In the Russian Federation, fundamental economical and political changes during the last decades have reinforced the problem of sustainable remediation of brownfields, especially in city areas. The main reasons for that are:

- Large space requirements for growing offices, housing, shopping and service facilities;
- Former industrial areas, which were closed for economic and/or environmental reasons, represent a serious hazard to the population due to pollution emissions and also a potentially good investment medium due to the increasing land cost (Sojref & Weinig, 2005).

3. Principles of risk assessment

Risk assessment of contaminated land serves two general purposes:

1. It is used to determine the *significance* of contamination at a site;
2. It may be used to determine the level of clean-up required for the intended use of the site.

Risk assessment as a methodology is not limited to the assessment of contaminated land but often used for other purposes, varying from prevention of pollution from new chemicals and processess, through reliability engineering of industrial activities and new technologies, to environmental impact assessment. Risk assessment of contaminated sites is somewhat different from risk assessment in other fields. The evaluation of risk from soil contamination is not usually a preventive approach; the source is already there. In principle this makes the assessment easier because claims about exposure can be verified at the site. In practice, however, this advantage is rather limited due to the complexity of the source, the difficulties of performing experiments and the need to predict *future* exposure. This predictive element means that there is much in common with risk assessment methods used in other fields.

The assessment of soil quality in general is based on the determination of the concentration of pollutants in soils. The estimated concentration is compared with specific threshold values and the degree of contamination is evaluated. An assessment of health risks from a soil contaminant may consider whether total exposure from several pathways exceeds a critical (tolerable or acceptable) intake level. The total intake is compared with an appropriate health criterion (tolerable daily intake, etc.) that represents the maximum acceptable level of exposure because of a critical effect on a target organ or metabolic pathway. Exposure in excess of this threshold is then considered to indicate that the soil contaminant poses a significant risk to human health. Thus, many regulatory bodies all over the world have developed or are considering the development of soil quality values.

Before proceeding there are some terms that require definition (some taken from Barlow & Philp, 2005). *Toxicity* is the potential of a material to produce injury in biological systems (usually human in contaminated land risk assessment, but not necessarily so). A *hazard* is the nature of the adverse effect posed by the toxic material. *Risk* is a combination of the hazardous properties of a material with the likelihood of it coming into contact with sensitive *receptors* under specific circumstances. Risk is therefore a statistical entity, and the term significance is important. In the UK, Part IIA of the Environment Act, 1995 (routinely called "Part 2A") defines contaminated land as land where it appears, by reason of substances in, on or under the land, that:

1. Significant harm is being caused or there is a significant possibility of such harm being caused (e.g. Evans et al., 2006), or;
2. Pollution of controlled waters is being, or is likely to be, caused (DETR, 2000).

Significance in this context is linked to (Cole & Jeffries, 2009):

*   The margin of exceedance;
*   The duration and frequency of exposure;
*   Other site-specific factors that the enforcing authority may wish to take into account.

A *receptor* is the biological entity which may be at risk, and is usually humans in contaminated land investigations. Children are normally identified as the most sensitive receptor (Jeffries & Martin, 2009) because their intakes of food, water, air and soil are greater per unit body weight than in adults. A *source* is the source of the contamination, and a *pathway* is the means by which the source contaminants reach the receptor, which is described by the *source-pathway-receptor* approach to risk assessment.

The underlying principle of site remediation is to eliminate or modify one or more of the above factors such that the risk is reduced to meet site-specific requirements. Under the current regime in the UK, a site is only designated contaminated if a significant *pollutant linkage* is established. That is, there must be present a source, a pathway and a receptor. If a source cannot be connected to a receptor, in the UK legal definition the site does not constitute contaminated land (Clifton et al., 1999). Some sites may have several such pollutant linkages (DEFRA and Environment Agency, 2004), and it is this type of linkage analysis that allows remedial design strategies to be defined that are realistic and cost-effective. This often results in a strategy less conservative than one based on multifunctionality.

4. The risk assessment process

There are four key steps in the process of assessing the risks associated with pollutant linkages.

1. *Hazard identification.* This is the stage at which the chemicals present on a site are anticipated, along with their characteristics, e.g. their concentrations, water solubility and toxicity. Due to the likelihood of many tens or even hundreds of potential contaminants being present at a site, the hazard identification stage usually focuses on known contaminants of concern.This would be typical of an oil-contaminated site, where the oil itself is composed of perhaps hundreds of individual compounds.
2. *Exposure assessment.* This is the estimation of pollutant dosages to receptors, based upon the use of the site and the conditions therein. There are multiple facets to these calculations. Among the factors to be considered are exposure duration and frequency, mean body weight and future population growth or decline.
3. *Toxicity assessment.* This is the acquisition of toxicity data, such as dose-response, and its evaluation for each contaminant for both carcinogens and non-carcinogens.
4. *Risk characterisation.* This is an assignment of the level of risk to each pollutant linkage. For many contaminated sites the best that can be reasonably expected at an initial desk-based stage is a qualitative risk estimate, such as insignificant, low, medium or high. The amount of data required for quantitative risk characterisation may be beyond all but the most rigorously characterised sites, such as national high priority sites.

4.1 Qualitative and quantitative risk assessment

The purpose of the qualitative risk assessment is to assign the significance or degree of real risk, as opposed to perceived risk. It is based on a systematic assessment of site-specific critical factors using professional judgement and expertise in addition to guidelines and standards. The causal chain of source-pathway-receptor is the basis. Formulation of the remedial objectives and strategy will essentially identify whether the source and/or pathway should be the focus of remedial objectives, or whether protection of the receptor is a more viable option.

In quantitative risk assessment the aim is to assign values for existing and future deleterious effects associated with exposure. It requires high quality data and is often applied when a site is suspected to pose an unacceptably high risk to human health. One of the reasons that quantified risk assessments are so data-intensive is that not only direct pathways need to be considered. Indirect contact can occur when contaminants are transported through soil, groundwater, surface water, uptake or adsorption by plants, dusts or aerosols. Current understanding of the complex interactions between chemicals in the subsurface is low. Also, most contaminated ground has previously been used for industrial or chemical works, and the presence of made ground and foundations usually causes large uncertainty in the various fate, attenuation and transport processes that affect the movement of contaminants (US EPA, 1996). Figure 2 gives an overview of the steps involved in quantitative risk assessment (adopted from DEFRA, 2002).

It should also be noted that for petrochemical- and crude oil-contaminated sites, quantitative risk assessment is made more challenging by the complexity of the contaminant mixture (Kuyukina et al., 2009) and the effects of weathering on the bioavailability of risk-critical compounds. It is common for high heterogeneity to exist in the distribution of hydrocarbon contaminants which impacts risk assessment results and the success of remediation actions. For heavier fraction hydrocarbons such as paraffines and polycyclic aromatic compounds (PAHs), losses due to biotic and abiotic weathering processes may result in compounds with increased hydrophobicity and recalcitrance (Pollard et al., 2004). These compositional changes dramatically affect the affinity of the weathered hydrocarbon pool for risk-critical compounds such as prior to, during and following a cleanup treatment.

As previously stated, the receptor is usually human and therefore the ultimate purpose of quantitative risk assessment is the protection of human health. The risk to human health posed by contaminants on a site is dependent on the concentration of the contaminant and the means of exposure, e.g. skin contact, inhalation, ingestion. Essentially, the exposure from a certain contaminant can be quantified from the following equation or permutations of it (Ferguson, 1996).

$$\text{Exposure} = \frac{(\text{Soil Intake Rate})(\text{Exposure Time})(\text{Resorption Rate})(\text{Contaminant Concentration})}{\text{BodyWeight}}$$

Where:

- Exposure or Absorbed Dose = Daily mass of contaminant absorbed per day, divided by the body weight of the receptor (mg per kg body weight per day);
- Soil Intake Rate = Daily amount of soil a receptor is exposed to (grams);

- Exposure Time = Number of days of exposure to the contaminant;
- Resorption rate = Toxicokinetics-based, empirical value quantifying the daily transfer of contaminants from the intake medium into the systemic circulation;
- Contaminant Concentration = Concentration of contaminant in the uptake medium (mg per gram of soil);
- Body Weight = Mass of receptor (kg).

An understanding of the fate and transport of contaminants is crucial if a meaningful risk assessment is to be obtained. This analysis can be very complicated, since the number and types of processes affecting contaminants during transport is governed by both inherent contaminant characteristics and environmental conditions. Understanding these complex dynamic processes requires a best approximation of the environmental chemistry of contaminants (e.g. biodegradability, hydrophobicity,) and the environment at the site (e.g. geology, geochemistry). At the heart of such matters is the concept of bioavailability.

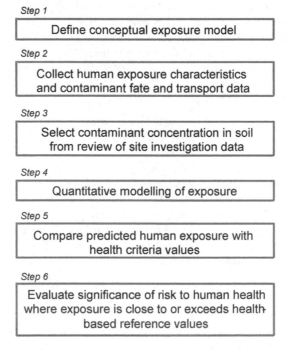

Fig. 2. Overview of the steps in quantitative site-specific risk assessment.

4.2 The conceptual model

A desk study should be undertaken for any given contaminated site to decide if enough information already exits to carry out a satisfactory risk assessment to a required degree of confidence. Such desk studies vary in investigative depth, but there are certain components that should be considered mandatory:

- The history of the site, including previous owners, occupiers and uses;

- A site visit, during which any visual evidence of potential contamination, site conditions and nearby features are recorded;
- Local geology and hydrogeology, including the presence and quality of groundwater and surface waters;
- The above and below ground layout of the site and its historical development;
- Any history of mining, including shafts and worked seams;
- Nearby waste disposal tips, abandoned pits and quarries;
- Information on previous investigations at the site;
- Processes used on the site including their locations, raw materials, products, waste and methods of disposal;
- Nearby sensitive receptors, e.g. water courses, houses, parks, areas of ecological sensitivity.

An interpretative desk study, i.e. one that not only provides factual data but also professional interpretation, will normally include a conceptual site model. This is a key component of the overall risk assessment process and is used as a tool to consider potential sources, pathways and potential receptors.

4.2.1 Generic conceptual model 1: Petrol station with operational spills and leaking underground storage tank

The following conceptual model (Fig. 3) and source-pathway-receptor matrix (Table 1) are from the Institute of Petroleum (1998). In the analysis of a petrol station with regular operational spills and leaking underground fuel storage tank (an exceedingly common occurrence; the US EPA estimated at one time over 200,000 in the US), there are many pathways by which the pollutants can reach receptors, and naturally there are several possible receptors. For example, an annual loss of petroleum products from operating petrol stations on the territory of Russia exceeds 160,000 tonnes, from which about 130,000 tonnes are lost during tank refuelling and fuel delivery. The impact of petrol stations on total air, soil and groundwater pollution of the world's large cities is estimated to be significant, causing a dramatic increase in health risks (Karakitsios et al., 2007). Estimated impacts of different sources in total evaporative emissions from petrol station operations are: filling underground fuel tank – 58%, underground fuel tank breathing and emptying – 3%, vehicle refuel operations – 37%, operational spillages – 2%.

A guide to good practice for development of conceptual models is available (McMahon et al., 2001). In the conceptual model, all possible combinations should be identified, but the matrix can be used to delineate which are the critical pathways and receptors, and which ones pose insignificant risk. The example is a great simplification. Each case will be site-specific with respect to geology, hydrogeology, geography (human population density is critical) and other factors. As a result, the source-pathway-receptor matrix can become complex. Once complete, the matrix saves time and effort as a number of insignificant risks can be identified and ignored.

Sufficient numbers of soil and groundwater samples should be collected in on-site and off-site areas affected by contamination originating from the petrol station. Laboratory analyses must be performed to provide the sample concentrations of individual contaminants, volatile organic compounds (VOCs): benzene, toluene, ethylbenzene, xylenes and methyl tert-butyl ether, semiVOCs: PAHs, which would be used as the bases for subsequent uniform and site-specific risk evaluations. The main chemicals of concern in the conceptual model of petrol station risk assessment are usually include benzene, toluene, ethylbenzene and methyl tert-butyl ether, among those benzene represents a highest hazard due to its toxic and cancerogenic effects. A comprehensive user guide for the human health risk assessment of petroleum releases at petrol stations is available (Joy & VanCantfort, 1999).

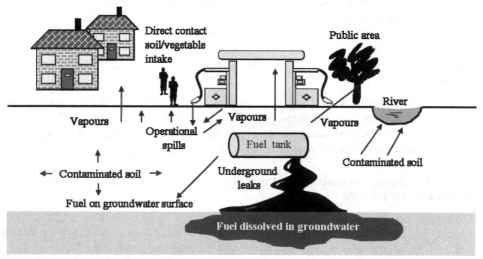

Fig. 3. Risk assessment for a petrol station with operational fuel spills and leaking underground storage tank.

4.2.2 Generic conceptual model 2: Crude oil spillage from a disrupted pipeline

Crude oil and petroleum products are widespread soil and groundwater pollutants resulting from spillage from the storage tanks and damaged pipelines. There are thousands of sites that have been seriously contaminated by petroleum products in oil-producing regions around the world (Etkin, 2001). In the US (Restrepo et al., 2009) crude oil is far and away the most frequently spilled hazardous liquid (39.4% of all cases, compared to gasoline, in second place, with 10.6%). Also the most frequent cause is easily demonstrated (a recent example – the Yellowstone River oil spill in Montana from the disrupted crude oil pipeline on July 1, 2011, when 1,000 barrels of crude oil went into the flood-swollen river); external and internal corrosion is the cause of 13.4% of all hazardous liquid spills, over three times higher than the next highest cause.

For Russia and the former Soviet Union (FSU), reliable data are difficult to find. In 2003, the World Bank published a report on pipeline failures in the countries of the former Soviet Union. The data search identified 113 major crude oil spill accidents during the period 1986–96 (inclusive). Just under 90% of these occurred in Russia. Whilst corrosion was still a major cause, there were double the number of spills caused by mechanical failure (UNDP/World Bank Energy Sector Management Assistance Programme, 2003). According to the Russian Federation State Environment Report, in 2003 losses of Russian oil and gas companies were approximately 3% from the product transported (Epifantsev & Shelupanov, 2011).

Primary source	Secondary source	Hazard	Transport mechanism	Pathway	Exposure medium	Receptor
Fuel tank	None	Dizziness, CNS depression potential carcinogenicity	Vapour, through air	Vapour inhalation	Air	Humans
Fuel tank	None	Vegetative die-back, leaf function damage	Vapour, through unsaturated zone	Vapour absorption	Air	Vegetation, adjacent trees
Fuel tank	None	Derogation of groundwater quality	Product loss and vertical migration to water table	Groundwater dissolution	Water	Groundwater aquifer
Fuel tank	None	Derogation of surface water quality	Product loss and dissolution in groundwater	Base flow and discharge to adjacent surface water body	Water	Adjacent river
Fuel tank	Contaminated soils	Dizziness, CNS depression	Vapour, through unsaturated zone	Vapour inhalation	Air	Humans (recreation)
Fuel tank	Contaminated soils	Skin irritation, contact dermatitis when extreme	Contact with contaminated soil	Dermal contact at surface	Soil	Humans (recreation)
Fuel tank	Contaminated soils	Flammability	Vapour, through unsaturated zone	Vapour build-up in basement	Air	Humans (residential)
Fuel tank	Contaminated soils	Flammability	Vapour, through unsaturated zone	Vapour build-up in basement	Air	Property
Fuel tank	Contaminated soils	CNS depression asphyxiation	Vapour, through unsaturated zone	Vapour build-up in basement	Air	Humans (residential)
Fuel tank	Contaminated soils	Derogation of surface water quality	Bulk fluid, through unsaturated zone	Free product flow to adjacent river	Water	Adjacent river
Fuel tank	Free product on water table	Derogation of soil quality	Evaporation to overlying soils	Vapour phase	Soil vapour	Soil
Fuel pump	None	Derogation of soil quality	Spillage and percolation through cracked hardstanding	Leaching	Soil	Soil
Fuel pump	None	Various, potential carcinogenicity	Vapour, through air	Inhalation	Air	Humans (customers)
Operational spills	None	Vegetative die-back	Vapour, through unsaturated zone	Vapour absorption	Soil gases	Home grown produce
Operational spills	None	Various, potential carcinogenicity	Vapour, through unsaturated zone	Consumption of contaminated produce	Vegetable produce	Humans (residential)
Operational spills	None	Various, potential carcinogenicity	Vapour, through air	Inhalation	Air	Humans (customers)
Operational spills	Contaminated soils	Dizziness, CNS depression	Vapour, through unsaturated zone	Vapour inhalation	Air	Humans (customers)

Table 1. Source-pathway-receptor for the petrol station with leaking underground storage tank

A conceptual model (Fig. 4) for the terrestrial oil-spillage from a disrupted pipeline occurring near a river can be used in the source-pathway-receptor risk assessments. It should be noted that the risk assessment of terrestrial oil-spills is much less explored compared to marine oil spills. However, it is an important field for development, especially if the pipelines are situated near, or are planned to cross rivers, lakes or other water bodies (Yang et al., 2010).

Oil types can differ in viscosity, volatility and toxicity. These three characteristics are very important when oil spills are being evaluated, because these parameters can influence the risk assessment results. For example, river water has a density of 1.0 g/cm^3 (compared to sea water having a density between 1.02 and 1.03 g/cm^3, depending on the salt concentration). This means that a heavy oil, with a density of 1.01 g/cm^3, would float in ocean water, but sink in a river, causing a severe problem of sediment oil contamination (Muijs & Jonker, 2011). Also, the oil can be moved with the current and, thus, spread a long distance from its origin. Some oil will evaporate, up to 50 percent of the volume. Natural physical, chemical and biological processes can cause the oil to weather, changing its characteristics (Malmquist et al., 2007). For example, emulsification leads to water-in-oil or oil-in-water stable mixtures that can persist for years.

Harmful effects of oil spills on natural environments have been extensively studied. However, only few studies so far have focused on the effect of oil exposure on human health (Aguilera et al., 2010). This supports the need for appropriate risk assessment methodology for human populations exposed to spilled oils, including the workers involved in the cleanup, in order to evaluate not only possible immediate consequences for their health but also the medium- and long-term effects, and the effectiveness of the protective devices used.

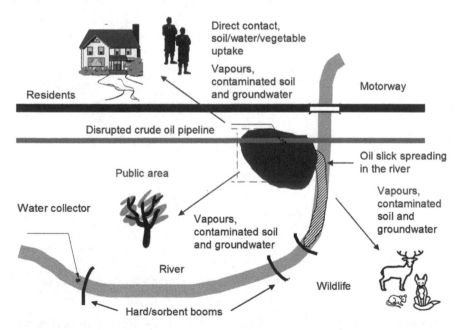

Fig. 4. Risk assessment for crude oil spillage from a disrupted pipeline.

5. The key role of soil guideline values and human health risk assessment

Soil Guideline Values (SGVs) are scientifically based generic assessment criteria that can be used to simplify the assessment of human health risks arising from long-term and on-site exposure to chemical contamination in soil. SGVs are guidelines on the level of long-term human exposure to individual chemicals in soil that, unless stated otherwise, are tolerable or pose a minimal risk to human health. They represent trigger, or intervention values – indicators that soil concentrations above this level are unacceptable.

Where representative soil concentrations of contaminants on a site are at or below the SGV, it can be assumed that it is very unlikely that a *significant possibility of significant harm* exists (DEFRA, 2008). Where representative soil concentrations of chemicals on a site exceed an SGV, further evaluation and assessment of the human health risks will normally be required to determine if a *significant possibility of significant harm* exists (Cole and Jeffries, 2009).

The situation in the US is similar. The US EPA Soil Screening Guidance (US EPA, 1996) is a tool to help standardise soil remediation at sites on the National Priority List (NPL). The outcome is soil screening levels (SSLs) for contaminants in soil that may be used for guidance purposes. SSLs are not national clean-up standards. At sites where contaminant concentrations fall below SSLs, no further action is warranted under the Comprehensive Environmental Response, Compensation and Liability Act (CERCLA). Generally, where contaminant concentrations exceed the SSLs, further investigation, but not necessarily clean-up, is warranted.

This resembles the 'trigger-action' approach and SSLs are risk-based concentrations derived from risk assessment procedures. However, the US EPA lists generic SSLs for 110 chemicals using default values that are conservative and likely to be protective for the majority of site conditions. It is noted that generic SSLs are not necessarily protective of all known human exposure pathways, reasonable land uses, or ecological threats.

The US soil screening is a seven-stage process:

1. Develop a Conceptual Site Model (CSM) based on historical records and available background;
2. Compare soil component of CSM to soil screening scenario;
3. Define data collection needs for soils to determine which site areas exceed SSLs;
4. Sample and analyse soil at site;
5. Derive site-specific SSLs if needed;
6. Compare site soil contaminant concentrations to calculate SSLs;
7. Decide how to address areas identified for further study.

Essentially, SSLs are risk-based concentrations derived from equations combining exposure assumptions with EPA toxicity data.

6. Risk assessment models

The nature of risk assessment i.e. gathering data, setting thresholds, statistical decision making and movement between different sections of a flow chart (see Fig. 1) make risk assessment procedures eminently suited to software modeling. Seventeen such human health risk assessment models (Table 2) have been identified recently in the literature (Cheng & Nathanail, 2009).

Many of the models have comparable approaches to assess health hazards arising from polluted soils. However, the input parameters and scenarios considered are different (Poggio et al., 2008). Results obtained with different methods are therefore often not comparable (European Commission, 2006). The Netherlands National Institute for Public Health and Environment (2002) recommended a toolbox on the European level including:

- Standardisation of the common elements;
- Definition of flexible elements to account for country/region specific (geographical, ethnological and political) peculiarities;
- Documentation on the sensitivity of calculated human exposure to the input parameters and guidelines on when and how to measure concentrations in the contact media;
- Information on the uncertainty/reliability of calculated human exposure.

Model name	Developer
CETOX-human	DHI Water and Environment and Danish Toxicological Centre, Denmark
CLEA 2002 • CLEA UK beta 2006 • CLEA UK 1.04	DEFRA and Environment Agency of England and Wales
CSOIL	RIVM, the Netherlands
JAGG	Denmark
LUR	LABEIN Technological Centre, Spain
No name given	INERIS, France
No name given	Kemarkta Konsult AB, Sweden
RBCA toolkit	ASTM, US
Report 4639	Sweden
RISC	Spence Engineering, US and BP, UK
RISC-HUMAN	Van Hall Instituut, the Netherlands
Risk Assistant	Hampshire Research Institute, US
ROME	ANPA, Italy
SFT 99:06	Norway
SNIFFER (1ST) SNIFFER (updated)	Land Quality Management, UK
UMS	Germany
Vlier-Humaan	VITO, Belgium

Table 2. Human health risk assessment models identified in the literature (Cheng & Nathanail, 2009)

7. Ecological risk assessment

Whilst human health protection is overwhelmingly the main objective of a risk assessment, the protection of ecosystems as a concept for risk assessment is a growing area. Ecosystem protection is based on the potential ecological risk to soils. It is intimately linked to the philosophy of soil protection, which derives from the realisation that soil is largely non-renewable, taking centuries to build a mere centimetre (European Commission, 2007), and yet it provides us with 95% of all human sustenance, and it can be destroyed very quickly. It is subject to erosion, loss of organic matter, salinisation, landslides, as well as contamination. Soil degradation is accelerating, with negative effects on human health, natural ecosystems and climate change, and the economy. In this context it is hardly surprising that the need to re-develop brownfield sites has acquired high priority. These concerns are also the drivers for the proposed Soil Framework Directive (Commission of the European Communities, 2006). If adopted, it would be the first pan-European, soil-specific legislation.

Although ecological risk assessment is a growth area, it is inherently more complex than human health risk assessment as it requires multispecies analysis (Smith et al., 2005). Equally the identification and assessment of the significant possibility of ecological harm is hard to diagnose, and subject to interpretation. Ecological risk assessments are still therefore at a development stage (Latawiec et al., 2011).

Soil protection values have been derived for different regulatory applications by different authorities. Criteria can be developed for three main applications (Fernández et al., 2006):

- *Screening values*: representing soil concentration levels that may cause potential ecological dysfunction and, therefore, if exceeded, will require a site-specific assessment;
- *Clean-up targets*: representing the objectives to be achieved in restoration processes. In some cases, these values represent a similar level of protection as the screening values, but in other regulations the decision is a balance of the restoration cost and the ecological benefit;
- *Intervention values*: representing concentrations which are indicative of seriously contaminated sites that require immediate clean-up or control actions.

Fernández et al. (2006) offered an overall process for characterisation of contaminated soils based on ecological risk assessment principles, which is based on both chemical and biological tools in the decision-making scheme to arrive at a classification of soils as low-risk or high-risk. The chemical and biological techniques involved, however, are not routine. They propose direct toxicity assessment on terrestrial plants, soil-dwelling invertebrates and soil microorganisms, and also toxicity of leachates to algae, *Daphnia* and fish. The approach also allows setting of Generic Soil Quality (GSQ) values for chemicals independently of the amount of available information.

8. Bioavailability, bioaccessibility and risk assessment

The concept of bioavailability fits perfectly with risk assessment. If a pollutant is present in soil or water but is not available to the biota, then it presents minimum risk (unless chemical conditions change that can subsequently increase the bioavailability). However, measurement of bioavailability is often a difficult task. For more than a decade regulators have directed concerted effort towards rationalisation of risk-based contaminated land policies recognizing bioavailability and bioaccessibility as concepts to be incorporated into risk assessments (Latawiec & Reid, 2009).

As soon as a pollutant reaches soil, the level that is biologically available may start to decline as the chemical becomes sequestered in the soil by sorbtion (Chung & Alexander, 1998). The current approach to exposure assessment commonly relies on the total concentration, but it will be clear that the level that is biologically available might not be related to this number (Tang et al., 1999). Bioavailability also has consequences for partitioning phenomena, biodegradability and toxicity that are described in some detail by Philp et al. (2005b). In addition, this bioavailable fraction is also dependent on the organism considered and the properties of the matrix in which the organism is exposed, and the effective exposure time. Indeed, there is no universally agreed definition of bioavailability (Peijnenburg et al., 2007).

Within the context of bioremediation, bioavailability can be regarded as the fraction of a given analyte that is in a form making it biodegradable (Semple et al., 2003). The bioaccessible fraction provides a reference not only to the amount of a substance readily available to an organism at a given instant (bioavailability) but also to the fraction potentially available over time (Semple et al., 2004). With respect to human health risk assessment, the bioaccessible fraction is defined as the fraction of a substance that is released from the soil, during such processes as digestion into solution making it available for absorption (measured *in vitro*), whilst bioavailability relates to the fraction that reaches the blood system via the gastrointestinal tract (Wragg and Cave, 2002).

Lack of statutory guidance has been cited as the main factor hampering the use of bioavailability and bioaccessibility data in regulatory decision-making (Latawiec et al., 2010). However progress is being made. The International Standards Organisation (ISO) has been working on guidance for the selection and application of methods to measure bioavailability in soil (ISO 17402, 2008). The draft was created as a response to an increasing demand for a validated pool of methods to be used in soil assessments and promotes the development and the introduction of the bioavailability concept in the context of specific site circumstances. The ISO guidance aims to specify boundary conditions and principles for the methods and is still under international panel consultation (Latawiec et al., 2011).

9. Risk based remedial design

SGVs are not derived explicitly to be used as remediation standards. The process for setting remedial objectives and standards for remediation is outlined in CLR 11 (DEFRA and Environment Agency, 2004). If risk assessment demonstrates unacceptable risks are associated with a site, then these need to be managed. At this stage Options Appraisal comes into play (see Fig. 1). There are three main stages of Options Appraisal (DEFRA and Environment Agency, 2004):

1. Identifying feasible remediation options for each relevant pollutant linkage;
2. Carrying out a detailed evaluation of feasible remediation options to identify the most appropriate option for any particular linkage;
3. Producing a remediation strategy that addresses all relevant pollutant linkages, where appropriate by combining remediation options.

The scope of this chapter is limited to the identification of feasible remediation options. The process starts with the setting of remediation objectives. The objectives will be site-specific, but general considerations are:

- Degree to which risks have to be reduced or controlled;
- Time frame for remediation. Often for developers, speed is of the essence to manage cash flow;
- Technical efficacy of the proposed technology(ies) used for remediation;
- Cost of the strategy;
- Public opinion;
- In future, sustainability issues are likely to become more important e.g. environmental impacts of the technologies.

Once the objectives are set, it is necessary to determine remediation criteria. Some of the quantitative measures that can be used are:

- SGVs;
- Site-specific assessment criteria resulting from the risk assessment;
- Engineering-based criteria e.g. the size and design of on-site biopiles in a bioremediation strategy.

An objective assessment of the advantages, limitations and costs of different remediation options should be done. The full range of legal, technical, stakeholder and commercial issues has to be taken into account.

In complicated cases, such as former gas works, petrol stations or oil-spill sites, no one remediation technique is likely to work over the whole site, and then a strategy involving a treatment train is required. At gas works, typically the site specific risk assessment process will identify PAHs, phenolics, ammonia and complex cyanides as the main drivers for remediation. But there will often be buried chemical storage tanks to be dealt with, concrete needing to be crushed, perhaps selective landfilling of untreatable waste and filling of excavated voids, as well as strategies for protection of controlled waters. Careful planning is required to make sure that each component activity is carried out smoothly, in the corrected sequence and effectively. A detailed account of the clean-up of the former gas works site at the location of the Millenium Dome is given by Barry (1999).

9.1 Eco-efficiency of remediation technologies

The application of eco-efficiency measures of remedial technologies is not common-place currently, but may become so in the future. Sending contaminated soil to landfill, for example, is incompatible with modern views on recycling, and is an inefficient way to use limited landfill availability. Landfill taxation is making this option more expensive, and a range of other treatment technologies has flourished. Some initial work on eco-efficiency of remedial technologies has been done by Sorvari et al. (2009) (Table 3).

Remediation method	Positive factors	Negative factors
Reactive barrier	Generally no need for removal of the barrier	Long-term operating costs, suitable only for some contaminants
Soil stabilisation, isolation	No need for soil removal; quick; can be economical	No removal of contaminants from environment; can be energy-intensive
Soil vapour extraction (SVE)	Generally cost-effective; low uncertainties in risk reduction	Suitable only for volatile contaminants; exhaust air needs to be treated
Incineration (mobile)	Effective contaminant removal	Flue gas treatment needed; energy-intensive; often needs fuel
Composting	Low cost; treated soil may be used for landscaping; no emissions requiring treatment	Suitable only for some organic contaminants; can be long duration; depends on contaminant concentrations
Landfill	Effective control of risks; soil can be used in daily cover	Not treatment; not suitable for re-use; becoming more expensive; not efficient use of landfill sites

Table 3. Eco-efficiency of selected land remediation technologies (modified from Sorvari et al., 2009)

10. Conclusion

The pollution of soil and groundwater caused by accidental petroleum bydrocarbon releases is a complex environmental problem in all industrialised countries. Early strategies for clean-up based on highly stringent standards resulted in unsustainable cost burdens. The shift to risk assessment and suitability for use decreases the cost burdens, and has also been a factor in the development of new remedial technologies, including bioremediation. Many countries have developed or are currently developing frameworks and procedures for assessing and managing the risks posed by contaminated sites. The objective of this chapter is to review the principles and procedures of risk assessment for terrestrial ecosystems exposed to petroleum contamination. Focus is made on the effects of petroleum hydrocarbon contamination on human health rather than on ecological risk evaluation.

Two examples of the generic conceptual model concidering a potential source-pathway-receptor chain are given for exceedingly common cases: a petrol station with regular operational spills and accidental crude oil spillage from a disrupted pipeline. These models can be used in *site-specific risk assessment* when local environmental conditions (e.g. geology, hydrogeology, geography, human population) are used to calibrate the model. It is widely accepted that models are powerful tools for integrating various elements in risk assessment such as site characterisation, contaminant fate and transport, exposure assessment and risk calculation. They are, however, abstract and simplified representations of complex systems and are based on numerous assumptions and approximations. It is therefore important that models are validated and tested in real-world situations, either as part of contaminated land risk assessments or in research projects.

The chapter also identifies a number of problems at a general methodological level, especially concerning bioavailability and bioaccessibility as concepts to be incorporated into risk assessments. Indeed, lack of fully appropriate assessment methods for these complex environmental processes clearly indicates further research needs in the context of current approaches for contaminated land risk assessment. In this way, further developing the risk assessment approach would provide a rational and objective basis for ecological priority setting and decision making, particularly for the selection of eco-efficient remedial technologies.

11. Acknowlegements

This work was partly supported by the Russian Ministry of Education and Sciences (contracts 16.518.11.7069; 16.513.12.3015) and the Russian Academy of Sciences Presidium program "MCB".

12. References

Aguilera, F., Mendez, J., Pasaro, E. & Laffon, B. (2010). Review on the effects of exposure to spilled oils on human health. *Journal of Applied Toxicology* 30, 291–301.

Alberini, A., Longo, A., Tonin S, Trombetta F & Turvan M (2005). The role of liability, regulation and economic incentives in brownfield remediation and redevelopment: evidence from surveys of developers. *Regional Science and Urban Economics* 35, 327–351.

Barlow LR & Philp JC (2005). Suspicions to solutions: characterizing contaminated land. In: *Bioremediation: Applied Microbial Solutions for Real-World Environmental Cleanup.* American Society of Microbiology, ISBN 1-55581-239-2.

Barry DL (1999). The Millennium Dome (Greenwich Millennium Experience Site) contamination remediation. *Land Contamination and Reclamation* 7, 177-190.

Catney P, Henneberry J, Meadowcroft J & Eiser JR (2006). Dealing with contaminated land in the UK through development managerialism. *Journal of Environmental Pollution Planning* 8, 331-356.

Cheng Y & Nathanail PC (2009). Generic Assessment Criteria for human health risk assessment of potentially contaminated land in China. *Science of the Total Environment* 408, 324–339.

Chung N & Alexander M (1998). Differences in sequestration and bioavailability of organic compounds aged in dissimilar soils. *Environmental Science and Technology* 32, 855–860.

Clifton A, Boyd M & Rhodes S (1999). Assessing the risks. *Land Contamination and Reclamation* 7, 27-32.

Cole S & Jeffries J (2009). Using Soil Guideline Values. Environment Agency, Bristol, UK, ISBN 978-1-84911-037-2.

Commission of the European Communities (2006). Proposal for a Directive of the European Parliament and of the Council establishing a framework for the protection of soil and amending Directive 2004/35/EC. COM(2006) 232 final.

Day SJ, Morse GK & Lester JN (1997). The cost effectiveness of contaminated land remediation strategies. *Science of The Total Environment* 201, 125-136.

DEFRA (2002). The contaminated land exposure assessment (CLEA) model: Technical basis and algorithms (R&D Publication CLR 10), Bristol, UK.

DEFRA and Environment Agency (2004). Model procedures for the management of land contamination. Contaminated Land Report 11. Environment Agency, Bristol, UK, ISBN 1844322955.

DEFRA (2008). Guidance on the legal definition of contaminated land, PB 13149. Department for Environment, Food and Rural Affairs, London UK.

de Sousa (2003). Turning brownfields into green space in the City of Toronto. *Landscape and Urban Planning* 62, 181-198.

DETR (2000). DETR Circular 2/2000, Contaminated Land: Implementation of Part IIA of the Environmental Protection Act 1990. HMSO, Norwich.

Diplock EE, Mardlin DP, Killham KS & Paton GI (2009). Predicting bioremediation of hydrocarbons: Laboratory to field scale. *Environmental Pollution* 157, 1831-1840.

Epifantsev B.N., Shelupanov A.A. (2011). Conception of interconnecting security system for trunk pipelines against intended threats. *Oil and Gas Business* 1, 20-34.

Etkin DS (2001). Analysis of oil spill trends in the United States and worldwide. *Proceedings of International Oil Spills Conference.* American Petroleum Institute Publication, Washington, pp 1291–1300.

European Commission (2006). Impact Assessment of the Thematic Strategy on Soil Protection. Document accompanying the Thematic Strategy for Soil Protection, Communication from the Commission to the Council, the European Parliament, the Economic and Social Committee and the Committee of the Regions. European Commission, Brussels.

European Commission (2007). Environment fact sheet: soil protection - a new policy for the EU. 10.06.2011. Available from http://ec.europa.eu/environment/pubs/pdf/factsheets/soil.pdf.

Evans J, Wood G & Miller A (2006). The risk assessment–policy gap: An example from the UK contaminated land regime. *Environment International* 32, 1066–1071.

Ferguson CC (1996). Assessing human health risks from exposure to contaminated land: a review of recent research. *Land Contamination and Reclamation* 4, 159-170.

Fernández MD, Vega MM & Tarazona JV (2006). Risk-based ecological soil quality criteria for the characterization of contaminated soils. Combination of chemical and biological tools. *Science of the Total Environment* 366, 466–484.

Honders A, Maas T & Gadella JM (2003). Ex-situ treatment of contaminated soil – the Dutch experience. Service Centrum Grond, The Hague, Netherlands. 10.06.2011. Available from http://www.scg.nl/SCG/files/treatment.pdf.

Institute of Petroleum (1998). Guidelines for the Investigation and Remediation of Retail Sites. Portland Press, Colchester, UK.

Ishizaka K & Tanaka M (2003). Resolving public conflict in site selection process - a risk communication approach. *Waste Management* 23, 385-396.

ISO 17402 (2008). Soil quality — Requirements and guidance for the selection and application of methods for the assessment of bioavailabilty of contaminants in soil and soil materials.

Jeffries J & Martin I (2009). Updated technical background to the CLEA model. Science Report SC050021/SR3. Environment Agency, Bristol, UK, ISBN 9-781-84432-856-7.

Joy, T., VanCantfort, C. (1999). User Guide For Risk Assessment of Petroleum Releases. West Virginia Division of Environmental Protection. Virginia, USA. 53 p.

Karakitsios, S.P., Delis, V.K., Kassomenos, P.A. & Pilidis G.A. (2007). Contribution to ambient benzene concentrations in the vicinity of petrol stations: Estimation of the associated health risk. *Atmospheric Environment* 41,1889–1902.

Kuyukina M.S., Ivshina I.B. & Peshkur T.A., Cunningham C.J. Risk based management and bioremediation of crude oil-contaminated site in cold climate. *Proceedings of IASTED International Conference on Environmental Management and Engineering.* ACTA Press, Anaheim, Calgary, Zurich, 2009. pp. 117-122. ISBN 978-0-88986-682-9.

Larson B, Avaliani S, Vincent J, Rosen S & Golub A (1999). The economics of air pollution health risks in Russia: a case-studyof Volgograd. *World Development* 10, 1803-1819.

Latawiec AE & Reid BJ (2009). Beyond contaminated land assessment: On costs and benefits of bioaccessibility prediction. *Environment International* 35, 911–919.

Latawiec AE, Simmons P & Reid BJ (2010). Decision-makers' perspectives on the use of bioaccessibility for risk-based regulation of contaminated land. *Environmental International* 36, 383–389.

Latawiec AE, Swindell AL, Simmons P & Reid BJ (2011). Bringing bioavailability into contaminated land decision making: the way forward? *Critical Reviews in Environmental Science and Technology* 41, 52-77.

Luo Q, Catney P & Lerner D (2009). Risk-based management of contaminated land in the UK: Lessons for China? *Journal of Environmental Management* 90, 1123-1134.

Malmquist, L.M.V., Olsen, R.R., Hansen, A.B., Andersen, O., Christensen, J.H. (2007). Assessment of oil weathering by gas chromatography–mass spectrometry, time warping and principal component analysis. *Journal of Chromatography A*, 1164 262–270.

Markus J & McBratney AB (2001). A review of the contamination of soil with lead. II. Spatial distribution and risk assessment of soil lead. *Environment International* 27, 399–411.

McMahon A, Heathcote J, Carey M & Erskine A (2001). Guide to good practice for the development of conceptual models and the selection and application of mathematical models of contaminant transport processes in the subsurface. National Groundwater & Contaminated Land Centre report NC/99/38/2, 121 pp.

Muijs, B., Jonker. M.T.O. (2011). Assessing the bioavailability of complex petroleum hydrocarbon mixtures in sediments. *Environmental Science and Technology* 45 3554–3561.

Peijnenburg WJGM, Zablotskaja M & Vijver MG (2007). Monitoring metals in terrestrial environments within a bioavailability framework and a focus on soil extraction. *Ecotoxicology and Environmental Safety* 67, 163-179.

Philp JC, Bamforth SM, Singleton I & Atlas RM (2005a). Environmental pollution and restoration: a role for bioremediation. In: *Bioremediation: Applied Microbial Solutions for Real-World Environmental Cleanup*. American Society of Microbiology, ISBN 1-55581-239-2.

Philp JC, Stainsby FM & Dunbar, SA (2005b). Partitioning and bioavailability. In: *Water Encyclopedia: Oceanography; Meteorology; Physics and Chemistry; Water Law; and Water History, Art, and Culture*. pp. 521-527. John Wiley & Sons, Inc. New Jersey.

Pickin J (2009). Australian landfill capacities into the future. Report prepared for the Department of the Environment, Water, Heritage and the Arts, Hyder Consulting Pty Ltd, report ABN 76 104 485 289.

Poggio L, Vrščaj B, Hepperle E, Schulin R & Marsan FA (2008). Introducing a method of human health risk evaluation for planning and soil quality management of heavy metal-polluted soils - An example from Grugliasco (Italy). *Landscape and Urban Planning* 88, 64–72.

Pollard, S.J.T., Hrudey S.E., Rawluck M., Fuhr B.J. (2004). Characterisation of weathered hydrocarbon wastes at contaminated sites by GC-simulated distillation and nitrous oxide chemical ionisation GC-MS, with implications for bioremediation. *Journal of Environmental Monitoring* 6, 713-718.

Restrepo CE, Simonoff JS & Zimmerman R (2009). Causes, cost consequences, and risk implications of accidents in US hazardous liquid pipeline infrastructure. *International Journal of Critical Infrastructure Protection* 2, 38-50.

Semple KT, Doick KJ, Jones KC, Burauel P, Craven A & Harms H (2004). Defining bioavailability and bioaccessibility of contaminated soil and sediment is complicated. *Environmental Science and Technology* 38, 228A–331A.

Semple KT, Morriss AWJ & Paton GI (2003). Bioavailability of hydrophobic organic contaminants in soils: fundamental concepts and techniques for analysis. *European Journal of Soil Science* 54, 809–818.

Smith R, Pollard SJT, Weeks JM & Nathanail PC (2005). Assessing significant harm to terrestrial ecosystems from contaminated land. *Soil Use and Management* 21, 527-540.

Sojref, D., Weinig, H.-G. (2005). Elaboration of a Guideline for Sustainable Regeneration of Industrial Brownfield Sites in the Russian Federation by example of St. Petersburg.

Final Report on the 5th EU framework program RESCUE-project "Best Practice Guidance for Sustainable Brownfield Regeneration".Werkstoffe & Technologien, Transfer & Consulting, Berlin, Germany.

Sorvari J, Antikainen R, Kosola M-L, Hokkanen P & Haavisto T (2009). Eco-efficiency in contaminated land management in Finland –barriers and development needs. *Journal of Environmental Management* 90, 1715–1727.

Tang J, Robertson BK & Alexander M (1999). Chemical-extraction methods to estimate bioavailability of DDT, DDE, and DDD in soil. *Environmental Science and Technology* 33, 4346–51.

The Netherlands National Institute for Public Health and the Environment (RIVM) (2002). Variation in Calculated Human Exposure. (RIVM report 711701030). National Institute for Public Health and the Environment, Bilthoven The Netherlands.

UNDP/World Bank Energy Sector Management Assistance Programme (ESMAP) (2003). Russia Pipeline Oil Spill Study. Report 60633.

US EPA (1996). *Soil Screening Guidance: User's Guide*. EPA/540/R-96/018.

US EPA (1997). Brownfields definition. US EPA Brownfields Homepage. www.epa.gov/brownfields.

Van Hees PAW, Elgh-Dalgren K, Engwall M & von Kronhelm T (2008). Re-cycling of remediated soil in Sweden: an environmental advantage? *Resources, Conservation and Recycling* 52, 1349-1361.

Wragg J & Cave MR (2002). In vitro methods for the measurements of the oral bioaccessibility of the selected metals and metalloids in soils: a critical review. R&D Technical Report P5-062/TR/01. ISBN 1857059867. Environment Agency, Bristol, UK.

Yang, S.-Z., Jin, H.-J., Yu. S.-P., Chen. Y.-C., Hao, J.-Q. & Zhai, Z.-Y. (2010). Environmental hazards and contingency plans along the proposed China–Russia Oil Pipeline route, Northeastern China. *Cold Regions Science and Technology* 64, 271–278.

Permissions

The contributors of this book come from diverse backgrounds, making this book a truly international effort. This book will bring forth new frontiers with its revolutionizing research information and detailed analysis of the nascent developments around the world.

We would like to thank Jatin Kumar Srivastava, for lending his expertise to make the book truly unique. He has played a crucial role in the development of this book. Without his invaluable contribution this book wouldn't have been possible. He has made vital efforts to compile up to date information on the varied aspects of this subject to make this book a valuable addition to the collection of many professionals and students.

This book was conceptualized with the vision of imparting up-to-date information and advanced data in this field. To ensure the same, a matchless editorial board was set up. Every individual on the board went through rigorous rounds of assessment to prove their worth. After which they invested a large part of their time researching and compiling the most relevant data for our readers. Conferences and sessions were held from time to time between the editorial board and the contributing authors to present the data in the most comprehensible form. The editorial team has worked tirelessly to provide valuable and valid information to help people across the globe.

Every chapter published in this book has been scrutinized by our experts. Their significance has been extensively debated. The topics covered herein carry significant findings which will fuel the growth of the discipline. They may even be implemented as practical applications or may be referred to as a beginning point for another development. Chapters in this book were first published by InTech; hereby published with permission under the Creative Commons Attribution License or equivalent.

The editorial board has been involved in producing this book since its inception. They have spent rigorous hours researching and exploring the diverse topics which have resulted in the successful publishing of this book. They have passed on their knowledge of decades through this book. To expedite this challenging task, the publisher supported the team at every step. A small team of assistant editors was also appointed to further simplify the editing procedure and attain best results for the readers.

Our editorial team has been hand-picked from every corner of the world. Their multi-ethnicity adds dynamic inputs to the discussions which result in innovative outcomes. These outcomes are then further discussed with the researchers and contributors who give their valuable feedback and opinion regarding the same. The feedback is then collaborated with the researches and they are edited in a comprehensive manner to aid the understanding of the subject.

Apart from the editorial board, the designing team has also invested a significant amount of their time in understanding the subject and creating the most relevant covers. They scrutinized every image to scout for the most suitable representation of the subject and create an appropriate cover for the book.

The publishing team has been involved in this book since its early stages. They were actively engaged in every process, be it collecting the data, connecting with the contributors or procuring relevant information. The team has been an ardent support to the editorial, designing and production team. Their endless efforts to recruit the best for this project, has resulted in the accomplishment of this book. They are a veteran in the field of academics and their pool of knowledge is as vast as their experience in printing. Their expertise and guidance has proved useful at every step. Their uncompromising quality standards have made this book an exceptional effort. Their encouragement from time to time has been an inspiration for everyone.

The publisher and the editorial board hope that this book will prove to be a valuable piece of knowledge for researchers, students, practitioners and scholars across the globe.

List of Contributors

Annalisa Pinsino
Dipartimento di Scienze e Tecnologie Molecolari e Biomolecolari (Sez. Biologia Cellulare), Università di Palermo, Italy
Istituto di Biomedicina e Immunologia Molecolare "Alberto Monroy" CNR, Palermo, Italy

Valeria Matranga
Dipartimento di Scienze e Tecnologie Molecolari e Biomolecolari (Sez. Biologia Cellulare), Università di Palermo, Italy

Maria Carmela Roccheri
Istituto di Biomedicina e Immunologia Molecolare "Alberto Monroy" CNR, Palermo, Italy

Jatin Srivastava
Department of Applied Sciences, Global Group of Institutions, Raebareli Road, Lucknow, UP, India

Harish Chandra
Department of Biotechnology, G. B. Pant Engineering College, Ghurdauri, Pauri Garhwal, Uttrakhand, India

Anant R. Nautiyal
High Altitude Plant Physiology Research Centre, H. N. B. Garhwal University, Srinagar Garhwal, Uttrakhand, India

Swinder J. S. Kalra
Department of Chemistry, Dayanand Anglo Vedic College, Civil-Lines, Kanpur, UP, India

Osiel González Dávila
The University of London – School of Oriental and African Studies (SOAS), United Kingdom

Juan Miguel Gómez-Bernal and Esther Aurora Ruíz-Huerta
Posgrado en Ciencias de la Tierra, Instituto de Geofísica, Universidad Nacional Autónoma de México (UNAM), México

Alexander Strezov
Institute for Nuclear Research & Nuclear Energy, Bulgaria

Masako Oishi, Yoshihiro Miwa and Nobuo Kurokawa
Department of Pharmacy, Osaka University Hospital, Japan

Shinichiro Maeda
Graduate School of Pharmaceutical Sciences, Osaka University, Japan
Department of Pharmacy, Osaka University Hospital, Japan

Flávio Aparecido Rodrigues
Universidade de Mogi das Cruzes, Laboratório de Materiais e Superfícies (LABMAR), Brazil

Solange Bosio Tedesco
Graduate Program in Agrobiology, Department of Biology, Universidade Federal de Santa Maria (UFSM), Santa Maria, RS Brazil
Graduate Program in Agronomy, UFSM, Santa Maria, RS Brazil

Haywood Dail Laughinghouse IV
National Museum of Natural History, Smithsonian Institution, Washington, DC, USA
College of Computer, Mathematical & Natural Sciences, University of Maryland, College Park, MD, USA

Pann Pann Chung and J. William O. Ballard
School of Biotechnology and Biomolecular Sciences, University of New South Wales

Ross V. Hyne
Centre for Ecotoxicology, NSW Office of Environment and Heritage, Australia

André Silva Almeida
Center for Parasite Immunology and Biology, CSPGF-INSA, Porto, Portugal
Center for the Study of Animal Science, CECA-ICETA, Porto, Portugal

Sónia Cristina Soares and Elisabete Magalhães Silva
Center for the Study of Animal Science, CECA-ICETA, Porto, Portugal

Maria Lurdes Delgado, António Oliveira Castro and José Manuel Correia da Costa
Center for Parasite Immunology and Biology, CSPGF-INSA, Porto, Portugal

Romualdo Rodrigues Menezes
Department of Materials Engineering, Federal University of Paraíba, Brazil

Lisiane Navarro L. Santana, Gelmires Araújo Neves and Heber Carlos Ferreira
Academic Unit of Materials Engineering, Federal University of Campina Grande, Brazil

M. S. Kuyukina and I. B. Ivshina
Institute of Ecology and Genetics of Microorganisms, Russian Academy of Sciences, Perm, Russia
Perm State University, Perm, Russia

J. C. Philp
Science and Technology Policy Division, Directorate for Science Technology and Industry, OECD, Paris, France

S. O. Makarov
Perm State University, Perm, Russia